BRITISH CENOZOIC FOSSILS

FOSSILS

(PALEOGENE, NEOGENE AND QUATERNARY)

edited by Jonathan Todd and Simon Parfitt

Natural History Museum, London

First published by the Natural History Museum, Cromwell Road, London SW7 5BD.
© The Trustees of the Natural History Museum, London 2017. All Rights Reserved.

First edition 1960, second edition 1963, third edition 1968, fourth edition 1971, fifth edition 1975, this sixth edition 2017, reprinted 2022.

ISBN 978 0 565 09305 1

Typeset by Fakenham Prepress Solutions, Fakenham, Norfolk NR21 8NN
Printed by 1010 Printing International Limited

Front cover: *Volutospina luctator* from Pl. 25, fig. 8.

Contents

Foreword

The 354 species of animals and plants illustrated in this handbook represent only a small proportion of the many thousands of species known from the British Cenozoic fossil record. For example, more than 800 species of mollusc alone have been recorded from a single formation of Eocene age at a single site – the Selsey Formation at Selsey Peninsula, West Sussex. Given the overwhelming species diversity present, this book aims to illustrate a wide range of the more commonly found macrofossils together with a selection of the more rarely found species. Molluscs predominate, reflecting their diversity and abundance in most strata. The selection of species to illustrate has been pragmatic, originally decided upon by the availability of high quality illustrations, many from Arthur G. Wrigley's publications on molluscs in *Proceedings of the Malacological Society*, supplemented by as wide coverage of other organisms as possible. The additional illustrations were made by Arthur Wrigley and the artists L. D. Buswell and L. Ripley. The original edition was planned by Arthur Wrigley, and largely prepared by Cyril P. Castell under the general supervision of Sir Leslie R. Cox.

This edition has been significantly revised and updated – even its title has changed to follow international stratigraphical recommendations. Over the past 40 years, our knowledge of the stratigraphy of the British Cenozoic has been revolutionized by improvements in biostratigraphy, based on microfossil groups that have greatly improved the dating and correlation with North Sea and northwestern European Cenozoic deposits. Other advances such as the development of sequence stratigraphy, magnetostratigraphy and radiometric dating have together led to better delimitation and correlation of onshore shallow marine strata and have permitted their increasingly precise correlation with non-marine strata. In the Pleistocene our knowledge of stratigraphy and dating has improved greatly with technological advances, for example, ice core and stable isotope studies. Consequently, two new stratigraphic charts are presented, for the Paleogene, and Neogene and Quaternary, respectively, that give a simplified overview of the current stratigraphic nomenclature of the most important British fossiliferous strata. The stratigraphic nomenclature followed is largely that used by the British Geological Survey as supplemented and updated by Chris King's 2016 work, *A revised correlation of Tertiary rocks in the British Isles and adjacent areas of NW Europe*. Where there are discrepancies between these we have mostly followed the latter but have also made a few minor modifications of our own. In the charts we have included only formations and have omitted members, with the exception of a few bearing marine faunas that have many species either confined to them or with ranges starting or terminating in them.

Our knowledge of the identity and relationships of the organisms illustrated in this handbook continues to improve with continued research and these improvements in taxonomy are reflected in updated names of both genera and species. Some of these names and identifications were long out-of-date and had not been changed since publication of the first edition in 1960. The result is that some familiar names, such as *Neptunea contraria* – formerly applied to the common sinistral whelk from the Red Crag Formation – have disappeared. In this case, it has been shown that the name correctly refers to a distinct living species in the Atlantic Ocean; the name of the extinct species is correctly *Neptunea angulata*. Elsewhere two illustrations have been removed as the specimens were erroneously identified. Perhaps most surprisingly, continuing research has revealed that at least six common species of Eocene molluscs figured herein have been misidentified over the past 56 years and actually represent new and currently unnamed species. Additionally, one species of shark has been shown to be synonymous with another and both are now united under a single name, *Physogaleus secundus*.

This edition has benefitted greatly from the expertise of a large number of scientists; Sandra Chapman, Jill Darrell, Peta Hayes, Jerry Hooker, Adrian Lister, Martin Munt, Lorna Steel, Malcolm Symonds, Steve Tracey and David Ward (all of the Natural History Museum, London), Didier Merle (Muséum national d'histoire naturelle, Paris), Graham Oliver (National Museum Wales), Jan Johan ter Poorten (Field Museum of Natural History, Chicago, USA), Richard Preece (University of Cambridge), and most especially Frank Wesselingh and Ronald Pouwer (Naturalis Biodiversity Center, Leiden, The Netherlands) and Peter Moerdijk (Middelburg, The Netherlands). To all, we extend our warmest thanks; any errors remain our own responsibility. Simon Parfitt is indebted to the Calleva Foundation for supporting his research.

Introduction

Over the past 66 million years of the Cenozoic Era many of today's dominant groups of macro-organisms evolved and diversified. On land the flowering plants (angiosperms) rapidly diversified together with most of the modern groups of vertebrates, including lizards, snakes and amphibians as well as birds and mammals – the post-Cretaceous radiation of the last giving rise to the popular term, 'Age of the Mammals'. In the marine realm too, many of the major groups of organisms that dominate today's seas came to prominence, including bony fish (teleosts), seafloor-dwelling molluscs (snails and bivalves), barnacles and crabs.

Onshore in Great Britain Paleogene (Paleocene, Eocene and Oligocene) strata are almost entirely restricted to southeastern England, Neogene strata (Pliocene only, the Miocene is almost entirely missing) are restricted to East Anglia (see map). Fossiliferous Quaternary (Pleistocene and Holocene) deposits are found throughout Great Britain but extensive marine and glacial strata are confined to East Anglia. In southeast England Paleogene rocks are represented by a remarkably complete series of sediments dating from Late Paleocene to Early Oligocene times (59–33 million years ago). At Whitecliff Bay, Isle of Wight 700 m (2,300 ft) thickness of sediment is preserved, the most extensive and complete exposure of this age in northwest Europe. Paleogene sediments were deposited in environments ranging from freshwater lakes to estuaries, lagoons and, most commonly, shallow marine seas. Just as in today's seas, the larger organisms bearing durable skeletons (and hence easily fossilized) are dominated both in species diversity and abundance by the shells of bottom-dwelling molluscs. Among vertebrates, fossilized teeth of sharks and rays may be common, as too are fish ear bones (otoliths). Fossils of other vertebrates tend to be much rarer, including those of mammals, which are usually represented by bones from carcasses that have floated out to sea before being buried. Fossil land plants are occasionally common and are represented largely by seeds, fruit and occasional leaves. In the Early Pliocene to Early Pleistocene age shallow marine sandy sediments ('Crags') of East Anglia a diverse range of molluscs and other invertebrates are preserved together with rarer terrestrial and marine vertebrate fossils.

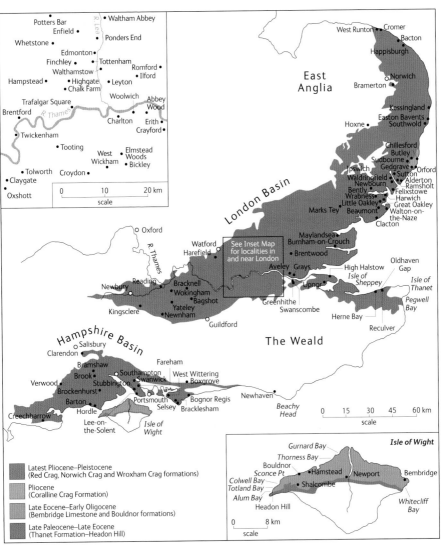

Distribution of significant Cenozoic (Paleogene, Neogene and
Quaternary) strata with the principal fossiliferous localities.

Later in the Pleistocene through to Holocene a wide range of freshwater and marine sediments, cave and river terrace deposits, together with glacial tills and outwash deposits, preserve a wide range of invertebrates and plants, and the teeth and antlers of large terrestrial mammals are frequently found.

British Cenozoic strata can be characterized as largely unlithified, siliciclastic sediments ranging from clays to coarse sands and conglomerates. Lithified freshwater limestones are usually very thin and comprise a small minority of the overall succession. These sediments form areas of low to moderate relief and are frequently overlain by Pleistocene and Holocene fluvial gravels. Natural exposures of unweathered sediments are mostly limited to coastal cliffs and foreshores with some small exposures inland in stream and river banks. Cliff exposures – particularly in the Hampshire Basin and East Anglia – are under threat both from ongoing erosion and also from coastal protection schemes that aim to mitigate this. The result has been great degradation and partial or total loss of some sites, for example at Bournemouth. Man-made exposures in sand and clay pits used to be common but the large majority have fallen into disuse or have been infilled. However, excellent temporary exposures have been made through road and rail construction and other civil construction works over the past few decades that have exposed a wide range of fossiliferous strata.

The Paleogene

The British onshore Paleogene deposits are generally shallow shelf to marginal marine sediments deposited on the southeastern marginal fringe of the North Sea Basin, a major depositional trough containing up to 3 km (2 miles) thickness of deep-water Cenozoic sediments. Onshore, deposits are largely confined to the synclinal London and Hampshire 'basins' (see Paleogene table pp.22–23). These are preservational remnants of a formerly continuous depositional area that was fragmented by tectonic uplift in the mid-Cenozoic. The North Sea Basin was usually connected to the Atlantic Ocean to the north and at various times a proto-English Channel connected it with the Atlantic to the west. Other seaways extended across present day Europe to the east and northeast. Throughout the Cenozoic, global climate change and the opening and closing of these seaways have controlled the overall nature of the marine faunas that compositionally range from cold temperate to tropical. Onshore sediments are characterized by frequent stratigraphic breaks and condensed horizons with their deposition controlled by both global sea-level change and patterns of local uplifting and subsidence. Marine transgressive horizons are often marked by a basal flint pebble lag.

Rocks of Early and Middle Paleocene age are missing due to major uplift of Britain triggered by a period of intense magmatic activity beneath Iceland associated with the opening of the North Atlantic Ocean. This coincided with a period of Alpine mountain building that produced basinal uplift throughout western Europe. As a consequence the oldest strata are of Late Paleocene age and these lie unconformably on Late Cretaceous chalks. The Thanet Formation (up to 30 m (100 ft) thick) and name-bearer of the Thanetian stage, comprises a sequence of fossiliferous silty clays to fine sands, sometimes siliceous, deposited in mid-shelf to nearshore marine environments. Their outcrop and subcrop include northern Kent and southern most East Anglia, and the London area west to Reading. At this time the proto-English Channel was closed and the faunas have a northerly aspect, are only moderately diverse and usually dominated by bivalves such as *Arctica* and *Astarte*. Excellent exposures are seen in the cliffs and foreshore around Herne Bay in northern Kent and also at Pegwell Bay. The overlying Upnor Formation (<15 m (<50 ft)) is a transgressive sequence of glauconitic sands and flint pebble beds that extends from northern Kent eastwards as far as the central Hampshire Basin. This unit was deposited in shallow marine conditions and is usually partly or wholly decalcified with oyster shells (*Ostrea bellovacina*) and sharks' teeth usually being the only macrofauna present. Unweathered horizons from excavations in the London area reveal a diverse molluscan fauna to have been originally present, indicating the presence of warmer Atlantic waters flowing through the Western Approaches. Exposures are present in the outlier at Newhaven, East Sussex as well as at quarries, including Charlton, Swanscombe and Lower Upnor, Kent.

In the latest Paleocene, tectonic uplift led to much of southern and eastern England forming a wide coastal plain on the margin of a North Sea that had become almost completely cut-off from the Atlantic due to extensive magmatic eruptions around Iceland and northern Britain. The earliest Eocene (early Ypresian stage) strata comprise the Woolwich and Reading formations that were deposited on this coastal plain during a short period (200,000 years) of higher temperatures – the Paleocene–Eocene Thermal Maximum (PETM). The Reading Formation (up to 27 m

(90 ft)) comprises mostly unfossiliferous fluvial sands, flint pebble beds and red-mottled clays and occasional lignites formed in freshwater lagoons subject to extensive soil-formation under a humid tropical climate. The Woolwich Formation (up to 15 m (50 ft)) was deposited concurrently and to the seaward of the Reading Formation with which it inter-fingers. It comprises interbedded sands, silts and grey clays with lignites and frequent shell beds deposited in muddy lagoons on the margins of a reduced salinity North Sea. Low diversity faunas are dominated by bivalves and gastropods. The Reading Formation is exposed in Alum and Whitecliff bays, Isle of Wight, Felpham in West Sussex – where an important lignite is preserved – and Redend Point, Studland Bay, Dorset. The Woolwich Formation is exposed in cliffs at Herne Bay and Newhaven, and quarry exposures at Charlton and Lower Upnor.

Incised into deep channels into the Woolwich Formation, the fine sands and flint pebble beds of the Blackheath Formation (<24 m (<80 ft)) represent shallow marine transgressive deposits that are restricted to small outcrops in southeast London and northwest Kent. Shell beds, such as those exposed in a small sand-pit at Abbey Wood, may be highly fossiliferous with a moderately diverse range of shallow marine molluscs intermixed with transported lagoonal taxa such as corbiculid bivalves. Here a rich and diverse fish, especially shark, fauna occurs with a much rarer but very diverse land mammal fauna. Although the succeeding Harwich Formation is relatively thin (<24 m (<80 ft)) it records an extensive marine transgressive system extending as far west as the western Hampshire Basin. The formation has a highly erosive base and comprises a geographically and stratigraphically complex suite of coastal marine deposits of silty clays, silts and fine glauconitic sands and channelled flint pebble conglomerates. In East Anglia more offshore silty clays occur that contain bands of volcanic ash. These ash bands were produced during an intensive phase of explosive volcanism in the Icelandic region 55 million years ago and ash can be traced from the Faroe Islands across the North Sea to northwestern Europe. More widely occurring sand and silty sand units may be highly fossiliferous and contain a low to moderately high diversity fauna of molluscs. Fossiliferous sands (Oldhaven Member) can be seen at Herne Bay and Lower Upnor, Kent and fossiliferous silts (Tilehurst and Swanscombe Members) are visible in a quarry at Harefield (Greater London). Exposures of the deeper water silty clays with ash bands occur on the foreshore at Harwich and in river cliffs at Wrabness, Essex (Orwell and

Wrabness members). They yield fish teeth, plants and occasional pockets of vertebrate remains, including a diverse bird fauna.

Extending across southern England, the London Clay Formation (Early Eocene, Ypresian) comprises a thick sequence (up to 150 m (490 ft)) of marine silty clays, becoming sandier to its base and top. The formation represents six major transgressive cycles and in the Hampshire Basin the regressive tops of these often comprise thick cross-bedded nearshore sands. Due to a rise in global sea-level, much of the formation in the east of the region was deposited in mid-shelf depths (>100 m (>330 ft)). In the lower part of the formation connection of the North Sea with the Atlantic through the Western Approaches was re-established and the molluscan fauna became more tropical in composition. In deeper water silty clays the fauna is usually sparse, though an extensive marine fauna and a highly diverse drifted terrestrial flora comprising tree logs, fruit and seeds is known from extensive coastal exposures, notably the cliffs and foreshore of the Isle of Sheppey, Kent. Here, too, a globally important vertebrate fauna is preserved, often in nodules, and comprises fish, turtles and birds. In contrast, the shallower water sandy clays of the Hampshire Basin contain more abundant molluscan faunas, including shallow water shell-beds but a notable pyritized terrestrial insect fauna is known from Bognor Regis, West Sussex. The uppermost part of the formation in the London and Essex area comprises shallow water interbedded sands and silts, sometimes containing a molluscan fauna. The London Clay used to be exploited extensively for brick-making with numerous exposures in man-made pits. These have all but disappeared over the past few decades but extensive coastal exposures exist at Bognor Regis, the Isle of Wight, northern Kent coast and in cliff and foreshore exposures in Essex, including Walton-on-the-Naze. In the Hampshire Basin the uppermost Ypresian is represented by a transgressive sequence (<53 m (<175 ft)) of shallow-water, dominantly lagoonal and intertidal laminated clays and sands with occasional lignites, containing a fully marine interval of fossiliferous glauconitic sandy clays and sands. This, the Wittering Formation, has a diverse but fragmentary fauna of shallow-water marine molluscs that remains poorly known. A rich shark fauna is present with occasional terrestrial vertebrates. Just as in the London Clay Formation, large fruits of *Nypa*, a tropical palm, may be found indicating a wet tropical climate and the presence of a coastal fringe of mangroves. Natural exposures are restricted to the cliffs and foreshore of Alum and

Whitecliff bays and the foreshore of Bracklesham Bay, West Sussex. In the western London Basin, the Wittering Formation is replaced by the Bagshot Formation (about 55 m (180 ft)), a series of poorly exposed fine to coarse cross-bedded or laminated sands and a unit of laminated silty clay with lignites, deposited in shoreface and marine lagoonal environments. Outliers of sandy units exist further east in north London, south Essex and the Isle of Sheppey that may contain a decalcified marine mollusc fauna.

The Middle Eocene commences with a transgressive suite of glauconitic silty sands and sands, the Earnley Formation (early Lutetian stage), up to 25 m (85 ft) thick. This formation was deposited in shallow tropical marine conditions and is highly fossiliferous in the Hampshire Basin. It is also preserved in the western part of the London Basin but is usually decalcified at the surface, though rich faunas were formerly obtained from gravel pits at Yateley, Surrey. The fauna comprises highly diverse assemblages of well-preserved marine molluscs that can be correlated across the region. Many of the taxa are wide-ranging throughout northwestern Europe and indicate open connections with the tropical Atlantic that persisted through the Lutetian. Colonial corals occur in the *Nummulites laevigatus* bed, a horizon typified by the abundant large tests of the benthic foraminifer now known as *Nummulites britannicus*. A rich shark and ray fauna is also present. Again, natural exposures are restricted to Bracklesham Bay and Whitecliff Bay, Isle of Wight. The Marsh Farm Formation (up to 20 m (65 ft)) overlies this in the Hampshire and western parts of the London Basin. It mostly comprises lagoonal thin-bedded clays and silts with glauconitic sand laminae, with occasional channelled marine pebbly sand and glauconitic silty sand horizons. The laminated clays are mostly unfossiliferous but the lagoonal channels and marine incursions can contain oyster beds and a rich restricted marine to fully marine molluscan fauna, sometimes with abundant fish teeth. These horizons may be seen at Whitecliff Bay and Bracklesham Bay when foreshore exposures allow. In the central and western Hampshire Basin and western London Basin fully marine conditions return with the transgression of the Selsey Formation sea. The Selsey Formation comprises glauconitic sandy clays to fine sands and was deposited in a tropical shallow shelf sea. The formation is highly fossiliferous throughout, containing a notably diverse tropical molluscan fauna (>800 species), with peak species diversity occurring in seagrass horizons dominated by tiny gastropods and bivalves. Sharks and teleost fish

– represented largely by otoliths – are diverse and abundant. Small natural fossiliferous exposures occur in streambanks in the New Forest, Hampshire and extensive foreshore outcrops may occur on the Selsey Peninsula and were formerly present at Lee-on-the-Solent. Cliff exposures at Whitecliff Bay are mostly weathered but occasional foreshore exposures can be highly fossiliferous. In Alum Bay, East Dorset and West Hampshire early Lutetian sediments are represented by a sequence up to 180 m (590 ft) thick, the Poole Formation. This formation consists of four depositional sequences, each comprising medium to coarse fluvial and estuarine sands followed by kaolin and organic-rich laminated clays deposited in floodplain and swamp environments. Coarser gravels are present westwards. In the Poole and Wareham district there has been extensive quarrying of the kaolinitic clays. This formation inter-fingers with the Wittering Formation at Alum Bay and its uppermost clay unit corresponds with the Marsh Farm Formation further eastwards. Two plant beds with leaf compressions are present at Alum Bay. Overlying the Poole Formation in the same western part of the Hampshire Basin, the Branksome Formation (70 m (230 ft)) comprises fine organic-rich sands, interlaminated sands and muds and organic laminated clays and occasional lignites deposited in estuarine environments but exposures on the mainland are now restricted. At Alum Bay these strata are largely represented by well-exposed shoreface marine sands that correlate with the Selsey Formation and may interfinger with the latter.

Towards the end of the Lutetian, localized uplift led to a regional unconformity with erosion of the topmost Selsey Formation sediments. A subsequent marine transgression deposited the Barton Clay Formation (late Lutetian to Bartonian age), a sequence of glauconitic shelly clayey sands, silts and silty clays (ca 50 m (ca 165 ft)) that is preserved in the central Hampshire Basin. This formation is famous worldwide for its diverse (many hundreds of species) and sometimes very well preserved shallow marine mollusc fauna, though a diverse range of marine organisms and occasional terrestrial mammals is present. Extensive cliff exposures are present from Highcliffe eastwards to Barton and smaller exposures exist at Alum and Whitecliff bays and in stream exposures in the New Forest. The lower part of the Barton Clay, termed the Elmore Member (up to 25 m (85 ft) thick), comprises silty clays and clayey silts with a highly fossiliferous basal shell bed. This member seems to transition westwards into decalcified fine to medium sands and flint cobble gravels (the Boscombe Sand Formation

– up to 27 m (90 ft)) in eastern Dorset. These sediments were deposited in marine shoreface and estuarine environments. Excellent exposures exist in cliffs at Hengistbury Head and Mudeford, as well as at Alum Bay. The Barton Clay is overlain by the Becton Formation, a sequence of fine silty sands and sands up to 70 m (230 ft) thick with a shallow marine to reduced salinity fauna, dominated by molluscs. A more clayey horizon with the abundant bivalve *Chama squamosa* and a diverse fully marine fauna is present at its base. Above this the strata contain more brackish but moderately diverse faunas. A brackish organic silty clay horizon occurs at Becton Bunny within the extensive Barton Cliff exposures. Other cliff exposures occur at Alum Bay and Whitecliff Bay; the last localized uplift and erosion of the nearby Sandown Pericline is recorded by the presence of reworked Early and Middle Eocene fossils and clasts. An isolated hill in east Dorset preserves the Creechbarrow Limestone Formation (2 m (7 ft)) at its summit. This thin freshwater tufaceous limestone contains a diverse mammal fauna that probably correlates in time with the mid-Barton Clay Formation.

The succeeding Headon Hill Formation of Late Eocene (Priabonian) age is a heterogeneous grouping of 11 units (members) of alternating and contrasting shelly clay, silt and sand units with a few limestones and lignites. It is up to 140 m (460 ft) thick where fully developed on the Isle of Wight and erosively overlies cross-bedded sands of the Becton Formation throughout the central Hampshire Basin. Some of the members occur throughout the outcrop area but five are more geographically restricted or lenticular. Depositional environments varied from marginal marine to freshwater lakes, and brackish and hypersaline lagoons, with common soils, developed on a wide coastal margin. This landscape probably extended across southern and eastern England, France, Belgium and the Netherlands. The Colwell Bay Member (up to 30 m (100 ft)), comprising shelly silt and silty clay, represents a marine transgression about 20 m (65 ft) above the base of the formation and contains a fairly diverse fauna with northerly affinities, suggesting that the North Sea Basin at this time lacked a southwestern outlet. Brackish to freshwater units contain a well-preserved molluscan fauna, but usually low in diversity, and some horizons contain abundant vertebrate material and a species-rich terrestrial mammal fauna. A large flora is known that indicates a significantly cooler climate than that of the Early Eocene. Extensive exposures are present in cliffs from Barton-on-Sea eastwards to Lepe Point, Hampshire as well as extensive coastal cliffs, especially in the southwestern Isle of Wight (Headon Hill, Colwell Bay) and in Whitecliff Bay. Small natural exposures are present in streambanks in the New Forest. The Bembridge Limestone Formation (<9 m (30 ft)), erosively overlies this unit on the Isle of Wight. It predominantly comprises freshwater limestones and lime-rich clays deposited in marshes and freshwater lakes. This formation is highly fossiliferous and contains an abundant freshwater gastropod fauna and an important terrestrial gastropod fauna is also present, particularly at Prospect Quarry, Shalcombe. Good exposures are otherwise present in foreshore ledges at Whitecliff Bay.

Above this on the Isle of Wight is the Bouldnor Formation, a thick series (up to 120 m (395 ft)) of brackish, fluvio-estuarine and freshwater silts and clays, frequently with carbonaceous or mottled muds, and occasional thin sands and limestones. The position of the Eocene–Oligocene boundary is disputed; it is currently thought to lie about 10 m (35 ft) above the base of the Hamstead Member of the Bouldnor Formation. Low diversity brackish and freshwater mollusc assemblages are mostly concentrated in seams throughout the formation. The basal unit of the lowest member, the Bembridge Marls, is a marginal marine oyster bed. The uppermost unit (Cranmore Member) is a transgressive horizon (9 m (30 ft)) series of fossiliferous shallow marine clays of lowermost Oligocene (Rupelian) age and is the youngest Paleogene stratum in the Hampshire Basin. Apart from molluscs, a large terrestrial flora and mammal fauna is present. A very thin (10 cm (4 in)) limestone occurs near the base of the formation that contains a highly diverse and exceptionally well-preserved insect fauna, the largest in the British Tertiary, together with plant macrofossils. This formation is exposed in cliff and foreshore sections at Whitecliff Bay, Hamstead Cliff and Bouldnor, Thorness Bay and Gurnard Bay. Strata of upper Oligocene age are almost absent onshore in the UK and there is a hiatus of nearly 30 million years before deposition of the oldest marine Neogene deposits.

About 20 isolated Paleogene sedimentary basins are present in the western part of Great Britain extending from Devon and Cornwall in the south to the Hebrides in the north. Relatively few are onshore and exposures are mostly poor. At Bovey Tracey, Devon, there is a narrow, deep basin containing up to 300 m (985 ft) thickness of dominantly kaolinitic clays and lignite beds (Bovey Tracey Formation) containing a macroflora of Early Oligocene (Rupelian) age. On the Island of Mull, off the west coast of Scotland, the Ardtun Conglomerate Member (6 m (20 ft)) is an interbedded fossiliferous claystone, sandstone and conglomerate occurring within the Staffa Lava Formation, the latter comprising 900 m (2,955 ft) of basaltic lava. Fossils are dominated by a macroflora largely comprising leaves and are of Middle Paleocene age.

The Neogene

Miocene strata are mostly absent in the southeastern North Sea Basin and are absent at outcrop in East Anglia. Here a major unconformity exists between the Early Eocene Harwich and London Clay formations and the overlying Pliocene or Pleistocene sediments. The Neogene marine deposits of East Anglia (see Neogene and Quaternary table pp.24–25) have long been known as 'Crags', a local term for shelly sands. They have a wide outcrop in northeast Essex, Suffolk and Norfolk and have good fossiliferous coastal exposures but are usually at least partly decalcified inland. The stratigraphic succession is incomplete and the preserved nearshore, shallow marine sediments are confined to those deposited during periods of high sea level.

The oldest deposits belong to the Coralline Crag Formation of late Early Pliocene (Zanclean) to earliest Late Pliocene (Piacenzian) age. This formation (up to 20 m (65 ft)) comprises marine silty calcareous and skeletal carbonate sand, fine- to medium-grained, with occasional silts. Coarser horizons may be partly decalcified and subsequently indurated by diagenetic calcite cement. The bioturbated to cross-bedded sands were deposited in inner to mid-shelf depths under a warm, Mediterranean climate. This formation is highly fossiliferous and contains a diverse invertebrate fauna dominated by molluscs but barnacles and large globose bryozoan colonies are also conspicuous. The basal transgressive horizon, contains a lag of pebbles, phosphatic nodules and sandstone cobbles ('boxstones') – the last referred to an otherwise eroded unit, the 'Trimley Sands'. This lag deposit is largely of Early Pliocene age but some of the large shark teeth present, including the extinct Giant White Shark *Carcharocles megalodon*, may be as old as the Middle Miocene. Fossiliferous exposures of the Coralline Crag may be found at Ramsholt Cliff and inland quarries at Sudbourne and Aldeburgh, all in Suffolk. Small pockets of sand containing ferruginous sandstone blocks are found in solution pipes in the Chalk of the North Downs of Kent. Now termed the Lenham Formation, these ex-situ deposits contain a mouldic mollusc fauna of latest Miocene or earliest Pliocene age. Exposures are currently very limited. At St Erth, Cornwall, there is a small outlier comprising a thin sequence (10 m (35 ft)) of highly fossiliferous shallow marine sands and clays – the St Erth Formation. Its highly diverse microfauna and molluscan macrofauna indicate deposition in warm waters in the latest Pliocene (Gelasian). No exposure exists today.

The Quaternary

The beginning of the Pleistocene is now taken at 2.58 million years ago, when major cycles of climate change commenced and ice sheets first formed on the northern hemisphere continents. Initially, cold episodes, during which mountain glaciers probably first appeared, were separated by shorter warm episodes every 40,000 years or so. This period of frequent but rather subdued climatic oscillations was followed about 750,000 years ago by more extreme climatic fluctuations marked by longer, intensely cold 'glacial' episodes separated by warm, but short-lived 'interglacials' at roughly 100,000-year intervals. During interglacials, the average annual temperature reached or even exceeded that of today. Under the influence of the regular pulses of climate change the vegetation shifted from open grassland to light woodland and finally mixed oak forest of the climatic optimum. The influence of these climatic and vegetational cycles can also be seen in the animal populations, which also changed in synchrony with the climatic oscillations. During warming phases, warmth-loving mammals (thermophiles) made their way north from glacial refugia, to be replaced during cooling phases by cold-adapted mammals that were capable of subsisting on grasses and other herbaceous plants growing under often harsh, subarctic climatic conditions.

The earliest Pleistocene deposits in Britain are confined primarily to the 'Crag Basin' located in the eastern parts of Suffolk and Norfolk. The basin contains a complex succession of marine sediments, comprising muds and sands, often with rich marine mollusc faunas. The Crag deposits also contain microfossils (foraminifers and pollen) that provide a fragmented record of Early Pleistocene climate. The oldest unit, the Red Crag Formation (50 m (165 ft)) comprises shelly medium to coarse cross-bedded glauconitic sands that are oxidized and highly ferruginous at outcrop. There is a rich and diverse invertebrate fauna dominated by molluscs and foraminifera that record a cooling trend marked by periods of warm-temperate and cool boreal conditions. The base of the formation is of latest Pliocene age, but the remainder is of Early Pleistocene (Gelasian) age. The base is highly erosive and locally deeply channelled and lies upon Eocene units or Coralline Crag Formation. A conglomeratic lag, the 'Suffolk Bone-Bed' may be present at its base. This yields reworked Mesozoic and Tertiary molluscs, Miocene shark teeth and Late Miocene (?) and Early Pliocene mammal teeth and bones. Excellent exposures of the Red Crag are seen in cliffs on the coast of north Essex and south Suffolk, including Bawdsey Cliff and Walton-on-the-Naze, as well as inland pits such as Waldringfield Heath, Suffolk.

The Norwich Crag Formation comprises up to 25 m (85 ft) of shelly fine to coarse glauconitic sand, sometimes cross-bedded, with lenticles of silty clay and gravel. It unconformably overlies the lithologically similar Red Crag Formation and it too was deposited in marginal to fully marine, tidally influenced conditions. This formation dates to the latest Gelasian (Early Pleistocene) age and records fluctuations from temperate to cold climate in its changing pollen and marine microfossil assemblages. A rich fauna of marine molluscs and other invertebrates is present. In addition, the Red Crag and Norwich Crag formations have yielded remains of whales and other cetaceans, but only sparse fossils of terrestrial mammals occur. The latter include southern elephant *Mammuthus meridionalis*, tapir *Tapirus arvernensis*, mastodon *Anancus arvernensis*, comb-antlered deer *Eucladoceros*, an extinct zebra-like horse and gazelle. Fossiliferous exposures of the Norwich Crag occur in rapidly eroding sea-cliffs such as Easton Bavents and Covehithe, Suffolk, as well as a number of inland pits, including those at Chillesford and Wangford, Suffolk.

Younger marine deposits of the Wroxham Crag Formation ('Weybourne Crag' of earlier authors) include marine and marginal marine interbedded sands, gravels, silts and clays up to 20 m (65 ft) thick with numerous 'exotic' clasts (e.g. Carboniferous chert, Mesozoic sandstones, North Wales volcanics). The molluscan fauna of the Wroxham Crag Formation is notable for the abundance of the bivalve *Macoma balthica*, a regional indicator species for a late Gelasian cold episode in the southern North Sea Basin. This formation spans the Early to early Middle Pleistocene (2.58–0.478 million years ago) of eastern Norfolk and northeastern Suffolk. Associated and interdigitated with this are estuarine, fluvial and floodplain deposits (up to 6 m (20 ft) thick) deposited by rivers draining large areas of northern and central England (Ancaster River and Bytham River) and the Proto-Thames, which reached the southern North Sea Embayment in the region of central and northern East Anglia. These deposits of organic muds, clays, silts and sands are known as the Cromer Forest-bed Formation and are exposed in rapidly eroding cliff sections along the coast from Weybourne (north Norfolk) to Kessingland (Suffolk). Sites such as West Runton in Norfolk (the type locality of the Cromerian Interglacial) are famous for the quantity and preservation of their fossils; these include a spectacular array of beetles, plant remains and freshwater and terrestrial molluscs, as well as mammals ranging in size from bats and shrews to rhinoceros and elephants. Recently the earliest traces of humans in northwestern Europe, represented by stone tools and footprints, were discovered at Happisburgh on the Norfolk coast in a horizon representing estuarine tidal mudflats dating to between 1 and 0.78 million years ago.

Younger uplifted marine deposits occur along the foot of the South Downs in West Sussex. The oldest of these raised beaches is found at about 40 m (130 ft) above present day sea-level. Dating to around 500,000 years ago, the famous archaeological site at Boxgrove has yielded Palaeolithic stone tools in abundance from shallow marine deposits and associated ponds and land surfaces. Characteristic implements include butchery tools known as handaxes. The associated mammalian fauna records a transition from an interglacial to cooler conditions, when early humans were exploiting large mammals for food and the ready supply of good-quality flint for toolmaking.

Between about 478,000 and 424,000 years ago, a period of intense cold (the Anglian Cold Stage) resulted in an ice sheet that spread from the Scottish Highlands and covered much of the British Isles as far south as the Severn Estuary and Hornchurch east of London. This ice sheet was responsible for interrupting well-established networks of eastward-flowing rivers, obliterating some and diverting the course of other rivers, such as the Thames. At its maximum extent, the British Ice Sheet coalesced with Scandinavian ice in the North Sea and trapped a huge glacial lake between the ice-front and a chalk ridge that then linked Britain with France. The formation of the Strait of Dover is the result of a catastrophic deluge from this glacially impounded lake. The glacial sediments deposited at this time include tills and outwash deposits (e.g. the Lowestoft and Happisburgh formations) that contain far-travelled, glacially transported erratics. The Happisburgh Formation of tills, sands and gravels with a high abundance of flint and quartzose clasts, is up to 20 m (65 ft) thick and lies erosively upon the Wroxham Crag and Cromer Forest-bed formations. It is exposed in cliffs in northeast Norfolk. A further set of glacial outwash sands and gravels and till (Corton Formation) overlies the Happisburgh Formation in northeast Norfolk. It includes fine-grained sediments (Corton Sand Member) that have been interpreted as marine or glaciomarine on the basis of their well-preserved marine molluscs and microfossils. An alternative explanation is that the fossils are derived from older sediments in the North Sea Basin, having been reworked and transported by glacial processes. This formation is truncated by the extensive chalky tills of the Lowestoft Formation (up to 60 m (195 ft) thick), which also comprises glacial outwash sands and gravels, silts and clays and this occurs extensively over East Anglia and further afield in the East Midlands.

The Anglian ice sheet had a significant influence on the formation of the modern British landscape in other ways. For instance, after the ice melted, new drainage patterns (e.g. Thames, Severn and Trent) were established, some of which preserve the most continuous successions of this age exposed on land. Gravels, sands and muds laid down by these rivers commonly contain abundant fossil plant and animal remains, including humans and their artefacts, representing the changing environments of Britain over the last few hundred thousand years.

A detailed picture of interglacial conditions can be reconstructed from numerous river, lake and cave sites of the Last (Ipswichian) Interglacial, which lasted from about 126,000 to 115,000 years ago. Famous sites of this age include Trafalgar Square in the middle of London and Barrington (near Cambridge), where building work and chalk quarries have uncovered highly fossiliferous deposits. The occurrence of warmth-loving plants (e.g. Montpellier maple, water chestnut) and animals (e.g. European pond terrapin, *Emys orbicularis*) suggest that summers were slightly warmer than now. Hippopotamus (*Hippopotamus amphibius*) is a characteristic mammal of this interglacial but strangely both humans and horse (*Equus ferus*) appear to have been absent in Britain at this time. Other members of the Ipswichian 'Hippopotamus fauna' include straight-tusked elephant (*Palaeoloxodon antiquus*), rhinoceros (*Stephanorhinus hemitoechus*), fallow deer (*Dama dama*), aurochs (*Bos primigenius*, the wild ancestor of domestic cattle), lion (*Panthera leo*) and spotted hyaena (*Crocuta crocuta*). Animals that failed to recolonize during the Ipswichian were probably restricted to southern refuges during the preceding cold stage;

they migrated north as the climate improved but were halted by the English Channel as sea level rose and flooded the land connection between Britain and the continent. Fossiliferous raised-beach deposits at Portland Bill, Selsey Bill and The Gower (Wales), provide evidence for a high sea-level coinciding with the Ipswichian climatic optimum.

Most British Pleistocene fossil localities date to the Last Cold Stage, known in Britain as the Devensian. As with earlier cold stages, the Devensian was not uniformly cold but punctuated by short-lived temperate interludes (interstadials). The changing plant and animal communities during the Devensian have been documented largely from work in caves and river deposits. Radiocarbon dating can be applied to the later part of the Devensian, leading to a reasonably complete understanding of faunal and floral history. Devensian mammal faunas are characterized by a typical cold-stage mixture of arctic (reindeer, *Rangifer tarandus*; arctic lemming, *Dicrostonyx torquatus*; Norway lemming, *Lemmus lemmus* and muskox, *Ovibos moschatus*) and steppe-dwelling species (Saiga antelope, *Saiga tatarica* and ground squirrel, *Spermophilus*), together with mega-herbivores tolerant of a rather harsh, subarctic climate. The latter group includes mammoth (*Mammuthus primigenius*), woolly rhinoceros (*Coelodonta antiquitatis*), giant deer (*Megaloceros giganteus*), and bison (*Bison priscus*) that disappeared from Britain around the end of the Last Cold Stage.

Some 11,700 years ago, the climate warmed very rapidly heralding the beginning of the present Holocene interglacial period. The complex interplay of rising sea level and rebounding of the land following the glacial retreat is recorded in deep sequences of stratified freshwater peats, soil horizons with trees ('submerged forests'), estuarine muds and marine sands found beneath the Wash/Fen Basin, the Somerset Levels and other coastal embayments and estuaries. These deposits have yielded the remains of animals such as bears, wolves, aurochs, wild boars, beavers, walruses and Dalmatian pelicans, as well as those of domesticated plants and animals that first arrived with Neolithic agriculturalists about 6,000 years ago. The globally more widespread and numerically ever greater human population, supported by farming and industrialization, is having an increasingly serious impact on global climate and the natural world, leading some geologists to propose a new geological epoch, the Anthropocene.

Stratigraphical tables of British Cenozoic formations

The following tables have been revised to follow current lithostratigraphical usage. The Paleogene table is based on the nomenclature and correlations adopted by the British Geological Survey as modified and revised by King (2016). We have reconciled nomenclatural differences in the Solent Group by following the former and have made our own modifications to the correlation of strata between the western and central Hampshire Basin. For the Neogene and Quaternary table we follow current British Geological Survey lithostratigraphical nomenclature for marine and glaciomarine strata and provide a simplified chronological framework for Quaternary non-marine deposits.

Paleogene

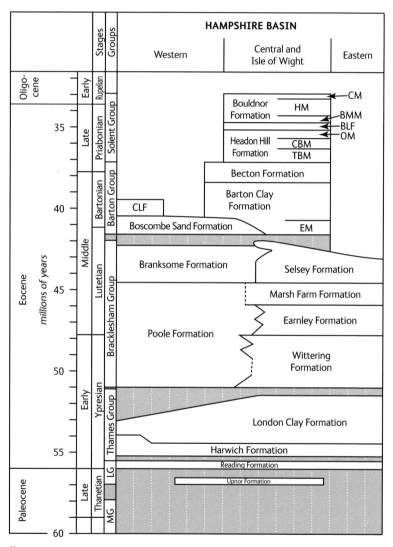

Key

BLF = Bembridge Limestone Formation
BMM = Bembridge Maris Member
CBM = Colwell Bay Member
CLF = Creechbarrow Limestone Formation

CM = Cranmore Member
EM = Elmore Member
HM = Hamstead Member
LG = Lambeth Group

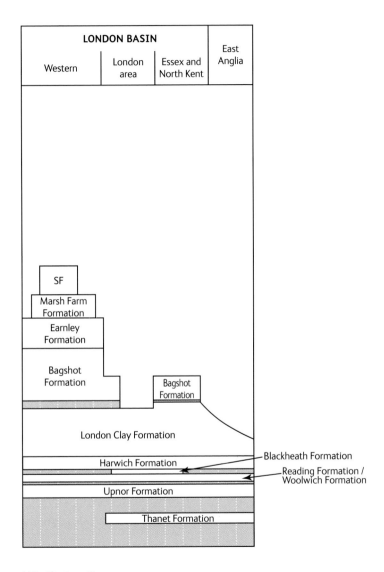

MG = Montrose Group
OM = Other members
SF = Selsey Formation
TBM = Totland Bay Member

Chart modified from C. King, 2016

Neogene and Quaternary

Age (Ma)	Period	Epoch			British/International Stages*	
0.0117		Holocene				
						Devensian (cold stage)
			Late	Tarantian*		
						Ipswichian (last) Interglacial
0.126						
			late Middle	Ionian*		Un-named cold stage
						'Aveley interglacial'
	Quaternary	Pleistocene				Un-named cold stage
						'Purfleet interglacial'
						Un-named cold stage
						Hoxnian Interglacial
			early Middle			Anglian (cold stage)
						'Cromerian Complex' and earlier warm and cold stages
0.781						
			Early		Calabrian*	
					Gelasian*	
2.588						
	Neogene	Pliocene	Late		Piacenzian*	
			Early		Zanclean*	

Key
Ma = millions of years

Marine	Non-marine
Clyde Valley Formation, Cambridgeshire Fens and Somerset Levels	
	Many glacial, cave and river terrace deposits, including Lea Valley, Upton Warren, Kent's Cavern ('Cave Earth'), Gough's Cave. Ballybetagh Mud, Castlepook Cave (Ireland)
Raised beaches of Portland Bill, Hope's Nose and Minchin Hole	Many cave and river deposits, including Kirkdale and Joint Mitnor caves, Bobbitshole, Barrington, Trafalgar Square
	Aveley, Crayford, Ilford, Pontnewydd Cave
	Purfleet, Grays Thurrock
	Swanscombe, Clacton, Hoxne, Marks Tey, Copford
	Corton, Happisburgh and Lowestoft formations
Goodwood-Slindon Raised Beach (Boxgrove)	Boxgrove, Westbury Cave, West Runton, Cromer Forest-bed Formation
Wroxham Crag Formation	
Norwich Crag Formation	Dove Holes (cave deposits)
Red Crag Formation	
Coralline Crag Formation	
'Trimley Sands' (Suffolk boxstones)	

The scientific names of fossils

The scientific name of a species is established by the publication with a description of its distinctive characters, in the case of fossils its morphology, and nowadays also with an illustration of the species. The scientist – known as a taxonomist – that describes a taxon (pl. taxa, a named taxonomic group) such as a genus or species is termed its author. The complete name of each species consists of two words which are usually Latin, derived from Latin, or treated as Latin. The first word is the name of the genus to which the species is assigned, and the second (the specific name) indicates the species. The family name of the author of a species is usually placed after the specific name; this gives a clue to where the literature description of a species is to be found. It also prevents confusion in those cases where the same name has mistakenly been established for two distinct taxa; these names are known as homonyms. When a number of authors sharing a family name have described species belonging to similar taxonomic groups, the authors initials may be added, for example J. Sowerby and J. de C. Sowerby, discriminate father and son. In the case of palaeobotany standardized initials are always used. If the species has been transferred to a different genus from that under which it was originally described, its author's name is placed in parentheses. If the specific name is an adjective, it must agree in gender with the generic name following the grammatical rules of Latin. Some specific names, however, are nouns in the genitive that do not change according to the gender of the genus.

Sometimes genera are subdivided into smaller morphological groups, termed subgenera. In these cases the subgeneric name is placed in parentheses between the generic and specific names. One subgenus will bear the same name as the original genus. Similarly, species may be divided into subspecies, that is, a group that differs morphologically from the original group the species was based upon, but to a lesser extent than that considered appropriate for recognition at species level. In such cases a Latin subspecific name is used, and this and its author's name follow the names already mentioned. The subspecies belonging to the group the species was initially based upon bears an identical specific and subspecific name. When names such as '*Striarca*' *wrigleyi* are given within single quotation marks it indicates that, in this case, *Striarca* is the closest generic name available but it is highly doubtful that the species is correctly placed in this genus. The qualifying term cf. (abbreviation of Latin *conferre*: compare) before a species name indicates the specimen is very similar but perhaps not identical to the named species. The term aff. (abbreviation of Latin *affinis*: having affinity with) indicates the specimen is distinct from, but closely related to, the named species.

The same taxon has sometimes been described by one or more taxonomists under different names, and in such cases the name first correctly published must be accepted. Discovery of earlier published names has been one reason for changes in nomenclature. A more important cause of changes in the names of organisms is because nomenclature is dependent upon taxonomy and classification and our knowledge of these is continually improving. Increased knowledge of taxa and their relationships may show that a species when first established was referred by its author to a genus which is now thought to be more distantly related. An example is the species from the Thanet Formation that was originally called *Pholadomya cuneata*, but which is now placed in a distinct genus, *Eutylus*, which belongs to a different family from *Pholadomya*. Moreover, one worker will treat a group of species as a distinct genus, while another will include the same group in a previously described genus. Similarly, one taxonomist will unite as a single species a series of specimens which another will consider to belong to two or more distinct species. The opinions of a wide range of taxonomic specialists have been sought to provide the most up-to-date names for the taxa illustrated in this latest edition. Other names that have been widely used in more recent literature and that differ in either the generic or the specific name, or in both are also given cited as synonyms (abbreviation, 'Syn.'). These names may be: 1) a true synonym, that is a different species name given subsequently to the same species; 2) an outdated combination of the species name with a generic name that is no longer considered correct; 3) an alternative opinion as to the correct genus and species combination; 4) a misidentification, in which case a phrase such as 'of British authors' or similar may be given after the name and its author; 5) a misspelling; 6) a homonym, or name that has previously been given to another species and which cannot now be used, in which case a phrase such as '*non* Linnaeus' is given that indicates the author of the earlier name.

Explanation of plates

The stratigraphic range given for each species is that currently known and applies to Great Britain only unless there is further comment. For Paleogene and marine Neogene species the range is given in terms of lithostratigraphic (rock-unit) formations, for terrestrial Pleistocene to Holocene species chronostratigraphic (time) ranges are given.

Two or more drawings bearing the same number and linked by a broken line are views of the same specimen.

The scientific names, quoted in square brackets after the abbreviation Syn. (= synonym), are other names that have been used, often incorrectly, for the species (see p.27).

Plate 1
Eocene and Oligocene Plants

1. *Anonaspermum rotundatum* E. Reid & M. Chandler. Seed. (× 1) London Clay Formation, Isle of Sheppey, Kent. RANGE: London Clay Formation.

2. *Iodes corniculata* E. Reid & M. Chandler. Part of fruit. (× 3) London Clay Formation, Isle of Sheppey, Kent. RANGE: London Clay Formation.

3. *Magnolia lobata* (Bowerbank). Internal cast of seed. (× 2) London Clay Formation, Isle of Sheppey, Kent. RANGE: London Clay Formation.

4, 5. *Oncoba variabilis* (Bowerbank). (× 1) 4, part of fruit; 5, segment of fruit with seeds. London Clay Formation, Isle of Sheppey, Kent. RANGE: London Clay Formation.

6. *Sabrenia chandlerae* M. Collinson. Seed (× 6) Headon Hill Formation: Totland Bay Member, Hordle, Hampshire. RANGE: Poole Formation–Bouldnor Formation: Hamstead Member. [Syn., *Brasenia ovula* (Brongniart).]

7, 8. *Wetherellia variabilis* Bowerbank. (× 1) 7, fruit; 8, segment of fruit. London Clay Formation, Isle of Sheppey, Kent. RANGE: London Clay Formation and Poole Formation.

9. *Stratiotes websteri* (Brongniart). Water Soldier seed. (× 3) Bovey Formation, Bovey Tracey, Devon. RANGE: Bouldnor Formation: Hamstead Member and Bovey Formation.

10. *Cinnamomum globulare* E. Reid & M. Chandler. Fruit. (× 1) London Clay Formation, Isle of Sheppey, Kent. RANGE: London Clay Formation.

11. *Gyrogona wrighti* (Salter). Gyrogonite. (× 30) Headon Hill Formation: Totland Bay Member, Hordle, Hampshire. RANGE: Headon Hill Formation–Bouldnor Formation: Hamstead Member. [Syn., *Chara medicaginula* Brongniart of authors.]

12. *Eomastixia rugosa* (Zenker). Part of fruit. (× 1) Headon Hill Formation: Totland Bay Member, Hordle, Hampshire. RANGE: Bagshot Formation–Headon Hill Formation. [Syn., *Eomastixia bilocularis* Chandler.]

13. *Hightea elliptica* Bowerbank. Worn fruit showing two seeds. (× 1½) London Clay Formation, Isle of Sheppey, Kent. RANGE: London Clay Formation.

14, 15. *Hightea turgida* Bowerbank. (× 1) 14, 15, parts of fruits. London Clay Formation, Isle of Sheppey, Kent. RANGE: London Clay Formation.

16. *Platycarya richardsoni* (Bowerbank). Fruiting head. (× 1) London Clay Formation, Swalecliffe, near Herne Bay, Kent. RANGE: London Clay Formation. [Syn., *Petrophiloides richardsoni.*]

Plate 1 continued opposite Plate 2

Plate 1

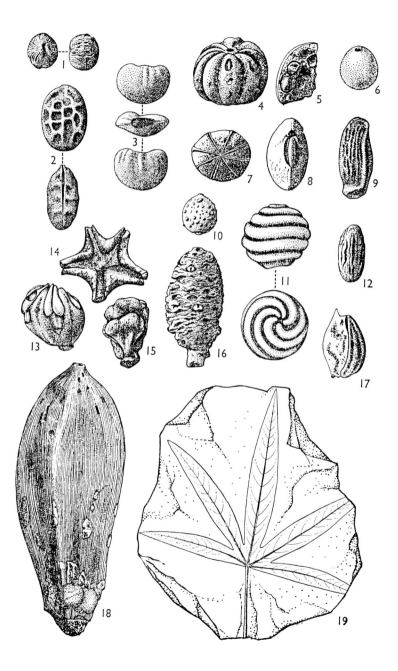

Plate 2
Eocene Foraminifera (Figs 1–6), Serpulid Worm Tube (Figs 7, 8) and Corals (Figs 9, 10)

1. *Palaeonummulites variolarius* (Lamarck). (× 15) Selsey Formation, Lee-on-the-Solent, Hampshire. RANGE: Marsh Farm Formation–Selsey Formation. [Syn., *Nummulites variolarius*.]

2. *Palaeonummulites prestwichianus* (Jones). (× 10) Barton Clay Formation: Elmore Member, Alum Bay, Isle of Wight. RANGE: Barton Clay Formation: Elmore Member [Syn., *Nummulites prestwichianus*, *Nummulites elegans* (J. de C. Sowerby).]

3. *Marginulina wetherelli* Jones. (× 15) London Clay Formation, Piccadilly, Greater London. RANGE: London Clay Formation.

4, 5. *Nummulites britannicus* Hantken in Hottinger & Schaub. (× 3) 4, part of surface removed to show whorls and septa; 5, vertical section. Earnley Formation, Southampton Docks, Hampshire. RANGE: Earnley Formation. [Syn., *Nummulites laevigatus* (Bruguière) of authors, *Nummulites lamarcki* d'Archiac & Haime.]

6. *Alveolina fusiformis* (J. de C. Sowerby). (× 5) Selsey Formation, Selsey, West Sussex. RANGE: Selsey Formation. [Syn., *Fasciolites fusiformis*, *Alveolina bosci* of authors, *Alveolina sabulosa* (Montfort).]

7, 8. *Sclerostyla mellevillei* (Nyst & le Hon). 7, tube with operculum in place (× 1); 8, operculum enlarged (× 3). Barton Clay Formation, Barton, Hampshire. RANGE: Earnley Formation?, Selsey Formation–Becton Formation.

9. *Turbinolia dixoni* Edwards & Haime. (× 2½) Earnley Formation, Bracklesham Bay, West Sussex. RANGE: Earnley Formation.

10. *Goniopora websteri* (Bowerbank). a (× 1); b (× 7½). Earnley Formation, Bracklesham Bay, West Sussex. RANGE: Earnley Formation.

Plate 1 continued

17. *Limnocarpus forbesi* (Heer). Part of fruit. (× 10) Headon Hill Formation: Totland Bay Member, Hordle, Hampshire. RANGE: Wittering Formation–Bouldnor Formation: Hamstead Member. [Syn., *Limnocarpus headonensis* (Gardner).]

18. *Nypa burtini* (Brongniart). Palm fruit. (× 1) (After Bowerbank.) London Clay Formation, Isle of Sheppey, Kent. RANGE: London Clay Formation–Selsey Formation and Branksome Formation. [Syn., *Nipa burtinii*.]

19. *Aralia sp.* Leaf. (× ½) Poole Formation, Alum Bay, Isle of Wight. RANGE: Poole Formation.

Plate 2

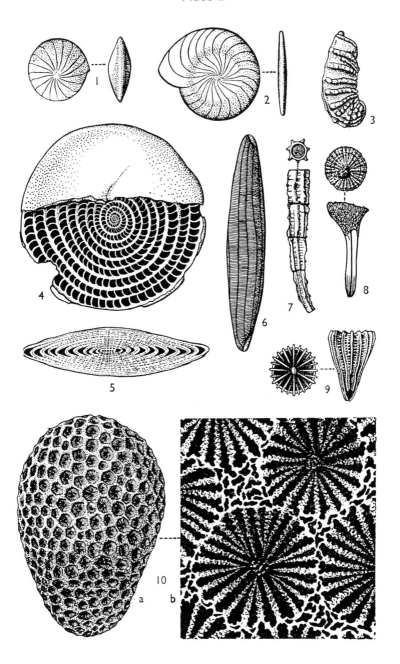

Plate 3

Eocene Brachiopods (Figs 1–5), Serpulid Worm Tubes (Figs 6–8) and Coral (Fig. 9)

1, 2. **Terebratula hantonensis** Muir-Wood. (1, × 1; 2, interior, × 1½)
London Clay Formation, Catisfield railway cutting, Fareham, Hampshire.
RANGE: London Clay Formation.

3. **Terebratulina wardenensis** Elliott. (× 1¼) London Clay Formation,
Isle of Sheppey, Kent. RANGE: London Clay Formation. [Syn., *Terebratulina striatula* of authors.]

4. **Lingula tenuis** J. Sowerby. (× 2) London Clay Formation, Bognor Regis,
West Sussex. RANGE: London Clay Formation–Earnley Formation.

5. **Discradisca ferroviae** (Muir-Wood). (× 3) Woolwich Formation,
Tooting, Greater London. RANGE: Woolwich Formation, Harwich Formation
(derived). [Syn. *Discinisca ferroviae*].

6. **Protula extensa** (Solander). (× 1) Barton Clay Formation, Barton,
Hampshire. RANGE: Selsey Formation–Becton Formation [Syn., *Serpula extensa*.]

7. **Ditrupa plana** (J. Sowerby). (× 1) London Clay Formation, Tolworth,
Greater London. RANGE: Harwich Formation–London Clay Formation.

8. **Rotularia bognoriensis** (Mantell). (× 1) London Clay Formation,
Bognor Regis, West Sussex. RANGE: London Clay Formation. [Syn.,
Vermetus bognoriensis.]

9. **Paracyathus crassus** Edwards & Haime. (× 10) London Clay Formation,
Clarendon, Wiltshire. RANGE: Harwich Formation–Barton Clay Formation.

Plate 3

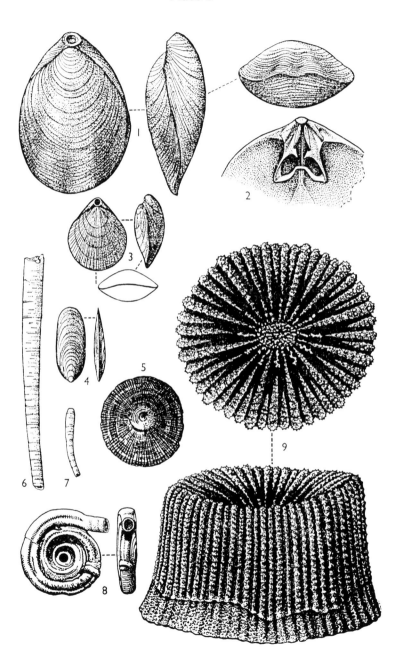

Plate 4
Eocene Arthropods (Figs 1–4, 6, 10–13) and Echinoderms (Figs 5, 7–9)

1–4. ***Arcoscalpellum quadratum*** (J. de C. Sowerby). Cirripede. 1, carina; 2, tergum; 3, scutum (all × 2); 4, complete capitulum (× 1½) London Clay Formation, Bognor Regis, West Sussex. RANGE: London Clay Formation.

5. ***Coulonia colei*** (Forbes)? Asteroid. (× 1) London Clay Formation, Isle of Sheppey, Kent. RANGE: London Clay Formation. [Syn., *Astropecten crispatus*, *Archastropecten crispatus* (Forbes).]

6. ***Zanthopsis leachi*** (Desmarest). Crab. (× 1) London Clay Formation, Chalk Farm, Greater London. RANGE: London Clay Formation.

7. ***Ophiura bognoriensis*** Rasmussen. Ophiuroid. (× 1½) London Clay, Bognor Regis, West Sussex. RANGE: London Clay Formation. [Syn., *Ophiura wetherelli* Forbes, in part.]

8, 9. ***Isselicrinus subbasaltiformis*** (Miller). Crinoid. Stem ossicles and stem. (× 2) London Clay Formation, Haverstock Hill, Greater London. RANGE: London Clay Formation.

10. ***Xanthilites bowerbanki*** Bell. Crab. (× 1) London Clay Formation, Isle of Sheppey, Kent. RANGE: London Clay Formation.

11. ***Basinotopus lamarckii*** (Desmarest). Crab. (× 1) London Clay Formation, Isle of Sheppey, Kent. RANGE: London Clay Formation. [Syn., *Dromilites lamarcki*.]

12. ***Hoploparia gammaroides*** M'Coy. Lobster. (× 1) London Clay Formation, Isle of Sheppey, Kent. RANGE: London Clay Formation.

13. ***Glyphithyreus wetherelli*** (Bell). Crab. (× 1) London Clay Formation, Isle of Sheppey, Kent. RANGE: London Clay Formation. [Syn., *Plagiolophus wetherelli*.]

Plate 4

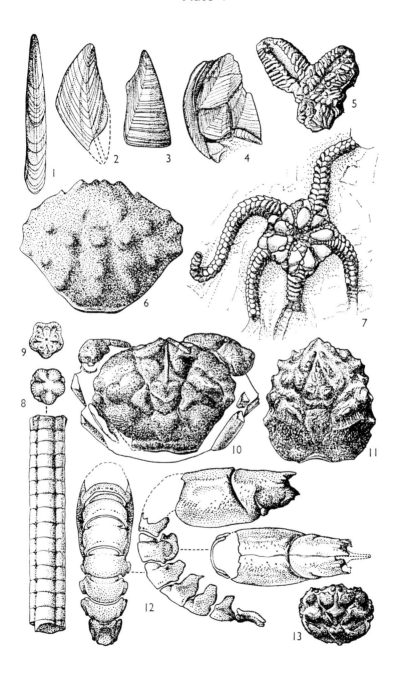

Plate 5
Eocene and Oligocene Bivalves

1, 2. **Nucula similis** J. Sowerby. (× 1½) Becton Formation. 1, Barton, Hampshire; 2, Alum Bay, Isle of Wight. RANGE: Barton Clay Formation–Becton Formation.

3, 4. **Nucula gracilenta** Wood. (× 1½) Blackheath Formation, Abbey Wood, Greater London. RANGE: Woolwich Formation?, Blackheath Formation–Harwich Formation. [Syn., *Nucula fragilis* Deshayes of authors.]

5, 6. **Ledina amygdaloides** (J. de C. Sowerby). (× 1½) London Clay Formation: 5, Potters Bar, Hertfordshire; 6, Chalk Farm, Greater London. RANGE: London Clay Formation. [Syn., *Nuculana amygdaloides.*]

7, 8. **Trinacria deltoidea laevigata** Mayer. (× 2) Headon Hill Formation: Colwell Bay Member, Headon Hill, Isle of Wight. RANGE: Becton Formation–Bouldnor Formation: Bembridge Marls Member. [Syn., *Trinacria curvirostris* (Cossmann), *Linter curvirostris.*]

9. **Glycymeris brevirostris** (J. de C. Sowerby). (× 1) London Clay Formation, Bognor Regis, West Sussex. RANGE: London Clay Formation.

10. **Glycymeris plumstediensis** (J. Sowerby). (× 1) Blackheath Formation, Abbey Wood, Greater London. RANGE: Thanet Formation–Harwich Formation. [Syn., *Pectunculus plumstediensis, Pectunculus plumsteadiensis.*]

Plate 5

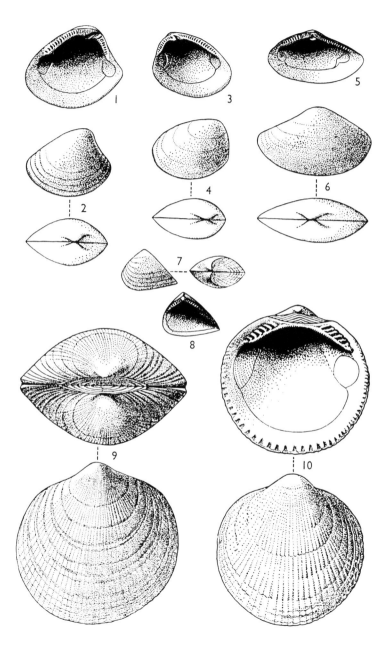

Plate 6
Eocene Bivalves

1, 2. **'Striarca' wrigleyi** (Curry). (× 1½) London Clay Formation, Highgate, Greater London. RANGE: London Clay Formation. [Syn.,'Glycymeris' wrigleyi, Glycymeris decussata (J. Sowerby non Linnaeus).]

3, 4. **Semimodiola elegans** (J. Sowerby). 3, internal mould (× 1); 4, shell (× 1½). London Clay Formation, Highgate, Greater London. RANGE: Blackheath Formation–Becton Formation. [Syn., Modiola elegans, Musculus elegans.]

5, 6. **Pectunculina scalaris** (J. de C. Sowerby). (× 1½) Becton Formation, Barton, Hampshire. RANGE: Barton Clay Formation–Becton Formation. [Syn., Limopsis scalaris.]

7. **Pteria media** (J. Sowerby). (× 1) London Clay Formation, Bracknell, Berkshire. RANGE: Harwich Formation–London Clay Formation. [Syn., Avicula media, Pterelectroma media.]

8, 9. **Glycymerita deleta** (Solander). (× 1) Becton Formation, Barton, Hampshire. RANGE: Barton Clay Formation–Becton Formation. [Syn., Glycymeris deleta.]

10. **Arca biangula** Lamarck. (× ¾) Earnley Formation, Bracklesham Bay, West Sussex. RANGE: Earnley Formation–Headon Hill Formation: Colwell Bay Member.

Plate 6

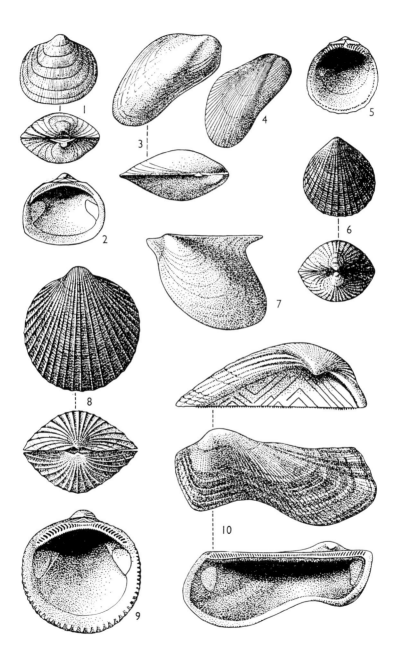

Plate 7
Paleocene and Eocene Bivalves

1. ***Atrina affinis*** (J. Sowerby). (× ½) London Clay Formation, Fareham, Hampshire. RANGE: London Clay Formation. [Syn., *Pinna affinis*.]

2. ***Mimachlamys recondita*** (Solander). (× 1½) Becton Formation?, Barton, Hampshire. RANGE: Barton Clay Formation–Becton Formation. [Syn., *Chlamys recondita*.]

3, 4. ***Cyclopecten duplicatus*** (J. de C. Sowerby). (× 1) 3, right valve; 4, left valve. London Clay Formation, Haverstock Hill, Greater London. RANGE: London Clay Formation. [Syn., *Pecten duplicatus, Chlamys woodi* Teppner.]

5. ***Lentipecten corneus*** (J. Sowerby). (× ¾) Selsey Formation, Bracklesham Bay, West Sussex. RANGE: London Clay Formation–Barton Clay Formation. [Syn., *Amusium corneum*.]

6, 7. ***Cucullaea decussata*** Parkinson. (× 1) Thanet Formation, Herne Bay, Kent. RANGE: Thanet Formation.

Plate 7

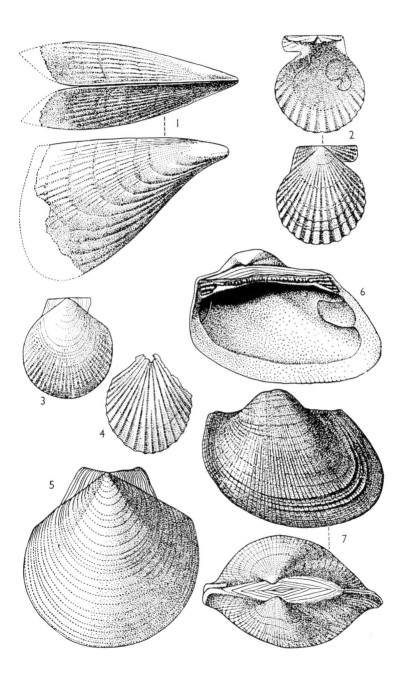

Plate 8
Eocene Bivalves

1, 2. **Striostrea velata** (Wood). (× 1) Headon Hill Formation: Colwell Bay Member, Colwell Bay, Isle of Wight. RANGE: Headon Hill Formation: Colwell Bay Member. [Syn., *Ostrea velata*.]

3. **Crassostrea tenera** (J. Sowerby). (× ½) Blackheath Formation, Upnor, Kent. RANGE: Woolwich Formation–Blackheath Formation [Syn., *Ostrea tenera*.]

4, 5. **Ostrea bellovacina** Lamarck. (× ½) 4, Blackheath Formation, Sundridge, Kent; 5, Woolwich Formation, Croydon, Greater London. RANGE: Upnor Formation–Harwich Formation.

Plate 8

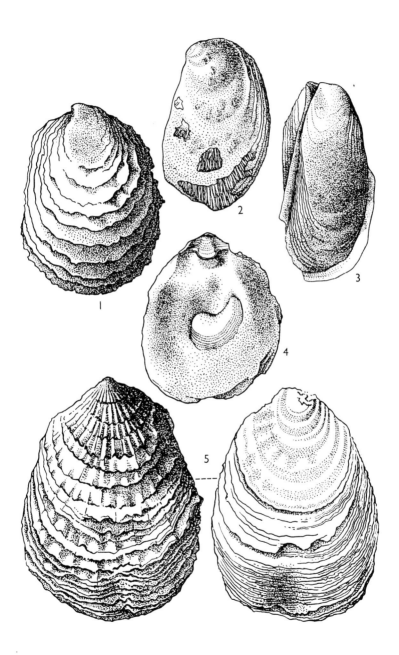

Plate 9
Paleocene and Eocene Bivalves

1. ***Cubitostrea plicata*** (Solander). (× 1) Barton Clay Formation, Barton, Hampshire. RANGE: Selsey Formation–Becton Formation. [Syn., *Ostrea flabellula* Lamarck of authors, *Ostrea plicata*.]

2, 3. ***Astarte* cf. *tenera*** Morris. (× 1) Thanet Formation, Herne Bay, Kent. RANGE: Thanet Formation.

4. ***Astarte filigera*** Wood. (× 1) London Clay Formation, Finchley, Greater London. RANGE: London Clay Formation.

5. ***Chattonia aequalis*** (Wood). (× 1) Becton Formation, Barton, Hampshire. RANGE: Becton Formation. [Syn. *Crassatella aequalis*.]

6. ***Glans oblonga*** (J. Sowerby). (× 1) Becton Formation, Barton, Hampshire. RANGE: Barton Clay Formation–Becton Formation. [Syn., *Venericardia oblonga, Cardita oblonga*.]

7, 8. ***Venericardia sulcata sulcata*** (Solander). (× 1) Barton Clay Formation, Barton, Hampshire. RANGE: Barton Clay Formation–Becton Formation. [Syn., *Cardiocardita sulcata*.]

9, 10. ***Bathytormus hemileius*** (Wood). (× 1) Selsey Formation, Brook, Hampshire. RANGE: Selsey Formation. [Syn. *Crassatella compressa* Lamarck of authors, *Crassatella compressa* var. *hemileia*.]

11, 12. ***Bathytormus sulcatus sulcatus*** (Solander). (× 1) Barton Clay Formation, Barton, Hampshire. RANGE: Barton Clay Formation–Becton Formation. [Syn. *Crassatella sulcata*.]

13, 14. ***Cyclocardia deltoidea*** (J. Sowerby). (× 1) Headon Hill Formation: Colwell Bay Member, Brockenhurst, Hampshire. RANGE: Headon Hill Formation: Colwell Bay Member. [Syn., *Venericor deltoidea, Venericardia deltoidea*.]

15. ***Venericor* sp.** (× 1) Wittering Formation, Whitecliff Bay, Isle of Wight. RANGE: Wittering Formation. [Syn., *Venericor planicosta suessoniensis* (d'Archiac) of authors.]

16. ***Crassatella sowerbyi*** Wood. (× 1) Selsey Formation, Stubbington, Hampshire. RANGE: Selsey Formation.

Plate 9

Plate 10
Eocene and Oligocene Bivalves

1–3. ***Polymesoda convexa convexa*** (Brongniart). (× 1) Bouldnor Formation, Hamstead, Isle of Wight. RANGE: Bouldnor Formation.

4, 5. ***Tellinocyclas tellinoides*** (Férussac). (× 1) Blackheath Formation, Abbey Wood, Greater London. RANGE: Woolwich Formation–Blackheath Formation. [Syn., *Corbicula tellinoides*.]

6–8. ***Loxoptychodon cuneiformis cuneiformis*** (J. Sowerby). (× 1) Woolwich Formation, Upnor, Kent. RANGE: Woolwich Formation– Blackheath Formation. [Syn., *Corbicula cuneiformis*.]

9, 10. ***Venericor planicosta*** (Lamarck). (× ½) Selsey Formation, Bracklesham Bay, West Sussex. RANGE: Earnley Formation–Barton Clay Formation: Elmore Member. [Syn., *Venericardia planicosta*, *Cardita planicosta*.]

11, 12. ***Polymesoda cordata*** (Morris). (× 1) Woolwich Formation, Charlton, Greater London. RANGE: Woolwich Formation. [Syn., *Corbicula cordata*.]

Plate 10

Plate 11
Paleocene, Eocene and Oligocene Bivalves

1. ***Polymesoda obovata*** (J. Sowerby). (× 1) Headon Hill Formation, Colwell Bay, Isle of Wight. RANGE: Headon Hill Formation–Bouldnor Formation. [Syn., *Corbicula obovata*.]

2, 3. ***Chama squamosa*** Solander. (× 1) Becton Formation, Barton, Hampshire. RANGE: Barton Clay Formation–Becton Formation.

4, 5. ***Arctica morrisi*** (J. de C. Sowerby). (× 1) Harwich Formation, High Halstow, Kent. RANGE: Thanet Formation–London Clay Formation.

6. ***Nemocardium (Nemocardium) nitens*** (J. Sowerby). (× 1½) London Clay Formation, Down Mill, Bracknell, Berkshire. RANGE: London Clay Formation.

7. ***Arctica planata*** (J. de C. Sowerby). (× 1) London Clay Formation, Portsmouth, Hampshire. RANGE: Harwich Formation–London Clay Formation.

Plate 11

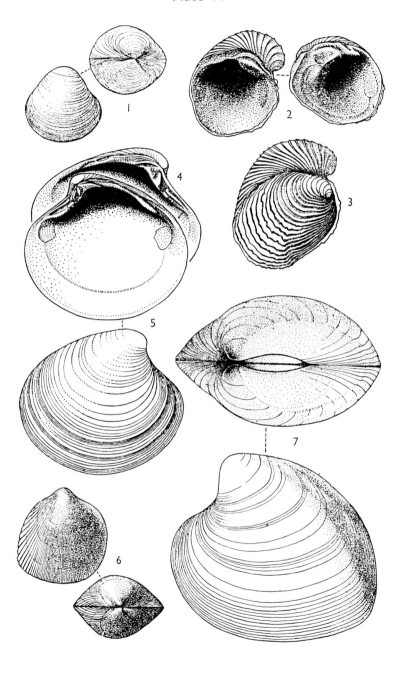

Plate 12
Paleocene and Eocene Bivalves

1–3. **Costacallista laevigata** (Lamarck). 1, (× ¾); 2, (× 1); 3, (× 1¼). Selsey Formation, Stubbington, Hampshire. RANGE: Earnley Formation–Headon Hill Formation: Colwell Bay Member. [Syn., *Macrocallista laevigata*.]

4, 5. **Nemocardium (Nemocardium) turgidum** (Solander). (× 1½) Barton Clay Formation, Barton, Hampshire. RANGE: Barton Clay Formation.

6–8. **Cordiopsis polytropa** (Anderson). (× ¾) Headon Hill Formation: Colwell Bay Member, Colwell Bay, Isle of Wight. RANGE: Headon Hill Formation: Colwell Bay Member. [Syn., *Cordiopsis incrassata* (J. Sowerby *non* Brocchi), *Sinodia suborbicularis* (Goldfuss) of English authors, *Pelycora* [sic] *suborbicularis*.]

9. **Axinus angulatus** J. Sowerby. (× 1) London Clay Formation, Lambeth Hill, Greater London. RANGE: London Clay Formation. [Syn., *Thyasira angulata*.]

10, 11. **Nemocardium (Arctopratulum) plumstedianum** (J. Sowerby). (× 1) Blackheath Formation, Abbey Wood, Greater London. RANGE: Thanet Formation–Harwich Formation. [Syn., *Nemocardium plumsteadianum* [sic].]

12. '**Thyasira**' **goodhalli** (J. de C. Sowerby). (× 1) London Clay Formation, Whetstone, Greater London. RANGE: London Clay Formation. [Syn., *Cryptodon goodhalli, Axinus goodhalli.*]

13. **Orthocardium porulosum porulosum** (Solander). (× 1) Barton Clay Formation, Barton, Hampshire. RANGE: Earnley Formation–Headon Hill Formation: Colwell Bay Member. [Syn., *Cardium porulosum, Vepricardium porulosum.*]

Plate 12

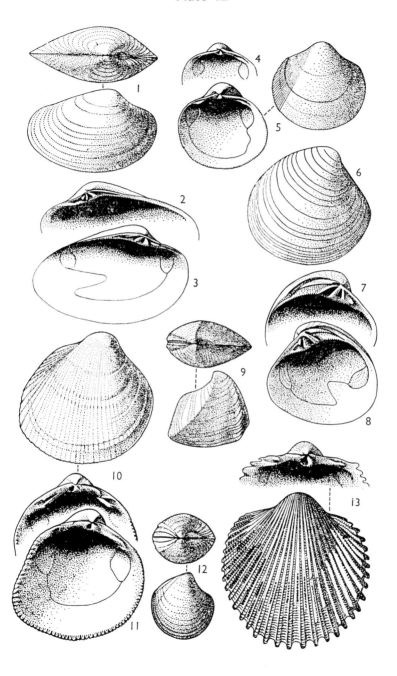

Plate 13
Paleocene and Eocene Bivalves

1–3. **Calpitaria tenuistriata** (J. de C. Sowerby). (× 1) London Clay Formation: 1, 3, Portsmouth, Hampshire; 2, Highgate, Greater London. RANGE: London Clay Formation. [Syn., *Pitar suessoniensis* (Watelet) of authors, *Pitar sulcatarius* (Deshayes) of English authors in part, *Calpitaria sulcataria*.]

4–6. **Microcallista proxima proxima** (Deshayes). (× 1) London Clay Formation, Portsmouth Dock, Hampshire. RANGE: London Clay Formation–Wittering Formation. [Syn., *Callista proxima*.]

7, 8. **Microcallista heberti belgica** (Vincent). (Exterior, × 1; hinge teeth, × 1½). Barton Clay Formation, Barton, Hampshire. RANGE: Barton Clay Formation. [Syn., *Callista heberti belgica*.]

9–11. **Dosiniopsis bellovacina** (Deshayes). (× 1) Thanet Formation, Herne Bay, Kent. RANGE: Thanet Formation–Harwich Formation. [Syn., *Dosiniopsis orbicularis* (Morris).]

12, 13. **Calpitaria vagniacarum** (Wrigley). (× 1) Harwich Formation, Swanscombe, Kent. RANGE: Harwich Formation. [Syn., *Pitar obliquus* (Deshayes) of authors, *Calpitaria obliqua*, *Pitar obliqua* var. *vagniacarum*, *Pitaria vagniacarum*.]

Plate 13

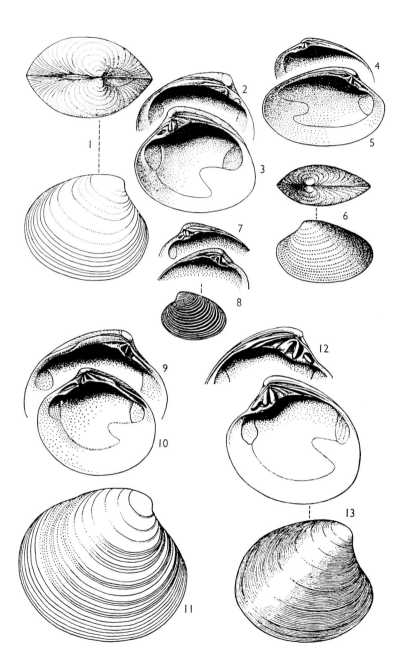

Plate 14
Paleocene and Eocene Bivalves

1. **Lentidium tawneyi** Curry. (× 3) Becton Formation, Long Mead End, Hampshire. RANGE: Becton Formation–Headon Hill Formation: Colwell Bay Member. [Syn., *Corbula nitida* J. Sowerby of authors.]

2. **Cultellus affinis** (J. Sowerby). (× 1) London Clay Formation, Highgate, Greater London. RANGE: Harwich Formation–London Clay Formation.

3–5. **Potamomya plana** (J. Sowerby). (× 1) Headon Hill Formation: Colwell Bay Member, Hordle, Hampshire. RANGE: Headon Hill Formation. [Syn., *Erodona plana*.]

6. **Abra splendens** (J. de C. Sowerby). (× 3) London Clay Formation, Highgate, Greater London. RANGE: London Clay Formation. [Syn., *Syndosmya splendens*.]

7–9. **Caryocorbula plicata** (Wrigley). (× 1½) Selsey Formation, Brook, Hampshire. RANGE: Selsey Formation–Barton Clay Formation: Elmore Member. [Syn., *Corbula plicata*.]

10. **Varicorbula sp.** (× 3) Selsey Formation, Brook, Hampshire. RANGE: Selsey Formation–Barton Clay Formation: Elmore Member. [Syn., *Corbula pisum* J. Sowerby of authors, *Corbula pisum* var. *wemmelensis* Vincent of authors, *Varicorbula* cf. *wemmelensis*.]

11–13. **Caestocorbula regulbiensis** (Morris). (× 2) Thanet Formation: 11, Herne Bay, Kent; 12, 13, Richborough, Kent. RANGE: Thanet Formation–Harwich Formation. [Syn., *Corbula regulbiensis*.]

14. **Caryocorbula cuspidata** (J. Sowerby). (× 2) Headon Hill Formation: Colwell Bay Member, Brockenhurst, Hampshire. RANGE: Becton Formation–Headon Hill Formation: Colwell Bay Member. [Syn. *Corbula cuspidata*.]

15. **Ficusocorbula ficus** (Solander). (× 1) Barton Clay Formation, Barton, Hampshire. RANGE: Selsey Formation–Becton Formation. [Syn., *Corbula ficus, Caestocorbula ficus*.]

16–18. **Bicorbula gallica** (Lamarck). (× 1) Earnley Formation?, Bracklesham Bay, West Sussex. RANGE: Earnley Formation–Barton Clay Formation: Elmore Member. [Syn., *Corbula gallica*.]

Plate 14

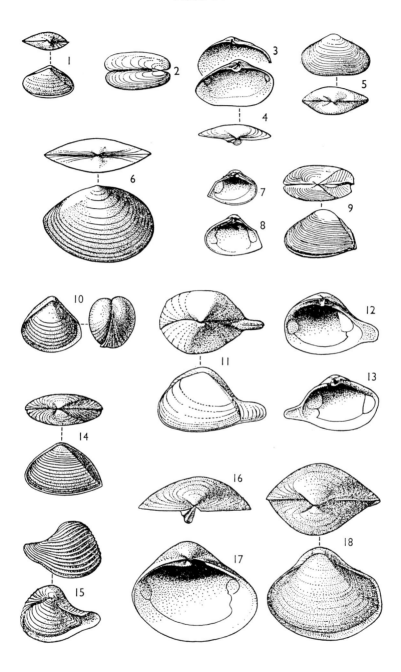

Plate 15
Paleocene and Eocene Bivalves

1. ***Thracia oblata*** (J. de C. Sowerby). (× ¾) Thanet Formation, Herne Bay, Kent. RANGE: Thanet Formation–London Clay Formation.

2–4. ***Psammotaea compressa*** (J. de C. Sowerby). (× 1) Barton Clay Formation, Barton, Hampshire. RANGE: Barton Clay Formation–Headon Hill Formation: Colwell Bay Member. [Syn., *Sanguinolaria compressa, Psammobia compressa.*]

5. ***Gari (Garum) edvardsii*** (Morris). (× ¾) Thanet Formation, Herne Bay, Kent. RANGE: Thanet Formation–London Clay Formation. [Syn., *Sanguinolaria edwardsi* of authors.]

6–8. ***Macrosolen hollowaysi*** (J. Sowerby). (× ¾) Earnley or Selsey Formation, Bracklesham, West Sussex. RANGE: Earnley Formation–Selsey Formation. [Syn., *Sanguinolaria hollowaysi.*]

9, 10. ***Gari (Garum) rude*** (Lamarck). (× ¾) Headon Hill Formation: Colwell Bay Member, Colwell Bay, Isle of Wight. RANGE: Becton Formation–Headon Hill Formation.

11, 12. ***Cyrtodaria rutupiensis*** (Morris). (× ¾) Thanet Formation, Herne Bay, Kent. RANGE: Thanet Formation–Harwich Formation.

Plate 15

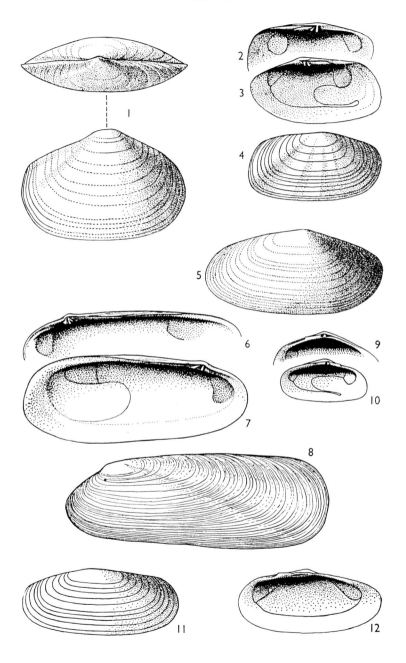

Plate 16
Paleocene and Eocene Bivalves

1–3. **_Teredina personata_** (Lamarck). Shipworm. London Clay Formation; 1, Highgate, Greater London (× 4); 2, ?Highgate, Greater London (× 1); 3, Tolworth, Greater London (× ¾). RANGE: Woolwich Formation–London Clay Formation.

4. **_Eutylus cuneatus_** (Morris). (× ¾) Thanet Formation, Herne Bay, Kent. RANGE: Thanet Formation. [Syn., _Pholadomya cuneata._]

5. **_Panopea intermedia_** (J. Sowerby). (× 1) London Clay Formation, Portsmouth, Hampshire. RANGE: Blackheath Formation–Headon Hill Formation: Colwell Bay Member.

6. **_Bucardiomya margaritacea_** (J. Sowerby). (× 1) London Clay Formation, Alum Bay, Isle of Wight. RANGE: Thanet Formation–London Clay Formation. [Syn. _Pholadomya margaritacea._]

Plate 16

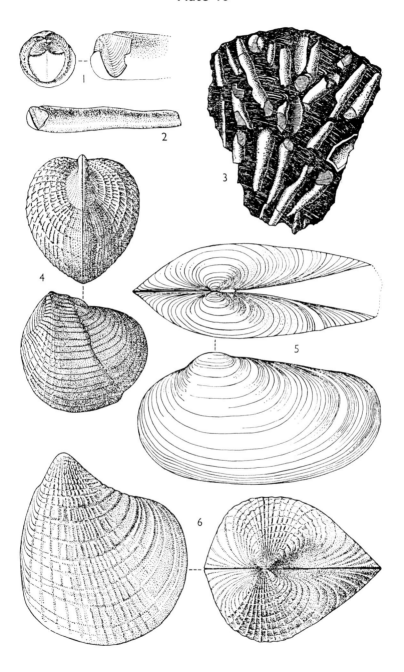

Plate 17
Eocene and Oligocene Gastropods

1. *Calliostoma nodulosum* (Solander). (× 1) Becton Formation?, Barton, Hampshire. RANGE: Barton Clay Formation–Becton Formation. [Syn., *Trochus nodulosus.*]

2, 3. *Pseudodostia aperta* (J. de C. Sowerby). (× 2) 2, operculum (a, outer and b, inner faces), Headon Hill Formation: Colwell Bay Member, Colwell Bay, Isle of Wight. RANGE: Headon Hill Formation: Colwell Bay Member. [Syn., *Neritina aperta, Theodoxus apertus.*]

4. *Clithon (Pictoneritina) planulatum* (Edwards). (× 2) Headon Hill Formation: Hatherwood Limestone Member, Headon Hill, Isle of Wight. RANGE: Headon Hill Formation: Hatherwood Limestone Member. [Syn., *Neritina concava* J. de C. Sowerby in part, *Theodoxus concavus.*]

5. *Rostreulima macrostoma* (Charlesworth). (× 7½) Barton Clay Formation, Barton, Hampshire. RANGE: Barton Clay Formation–Becton Formation. [Syn., *Eulima macrostoma.*]

6. *Acrilla reticulata* (Solander). (× 1½) Barton Clay Formation, Barton, Hampshire. RANGE: Selsey Formation–Barton Clay Formation. [Syn., *Epitonium reticulatum, Amaea reticulata.*]

7. *Granosolarium pulchrum* (J. de C. Sowerby). (a–c, × 1; d, sculpture of base, × 3; e, sculpture of spire, × 3) London Clay Formation, Highgate, Greater London. RANGE: London Clay Formation–Selsey Formation. [Syn., *Solariaxis pulcher, Stellaxis pulcher.*]

8, 9. *Polygireulima polygyra* (Charlesworth). (8, shell, × 7½; 9, apex, much enlarged.) Barton Clay Formation, Barton, Hampshire. RANGE: Selsey Formation–Barton Clay Formation,. [Syn., *Eulima polygyra, Polygyreulima* [sic] *polygyra, Eulima politissima* Newton.]

10. *Sigapatella aperta* (Solander). (× 1) Barton Clay Formation, Barton, Hampshire. RANGE: Selsey Formation–Headon Hill Formation: Colwell Bay Member. [Syn. *Calyptraea aperta.*]

11. *Nipteraxis bonneti* (Cossmann). (× 1½) Barton Clay Formation, Barton, Hampshire. RANGE: Barton Clay Formation–Becton Formation. [Syn., *Solariaxis bonneti, Architectonica bonneti, Solarium plicatum* of English authors.]

12. *Nystia duchasteli* (Nyst). (x. 2) Bouldnor Formation, Hamstead, Isle of Wight. RANGE: Bouldnor Formation. [Syn., *Tomichia duchasteli.*]

13. *Viviparus angulosus* (J. Sowerby). (× 1) Bembridge Limestone Formation, Sconce, Isle of Wight. RANGE: Headon Hill Formation–Bouldnor Formation: Bembridge Marls Member.

Plate 17 continued opposite Plate 18

Plate 17

Plate 18
Paleocene and Eocene Gastropods

1. **Sigatica sp.** (× 1) Blackheath Formation, Bickley, Greater London. RANGE: Blackheath Formation–Harwich Formation. [Syn., *Sigatica abducta* (Deshayes) of English authors.]

2. **Euspira glaucinoides** (J. Sowerby). (× 1) London Clay Formation, Clarendon, Wiltshire. RANGE: Blackheath Formation–London Clay Formation. [Syn., *Natica glaucinoides.*]

3. **Euspira bartonensis** Wrigley. (× 1) Barton Clay Formation, Barton, Hampshire. RANGE: Barton Clay Formation–Becton Formation.

4. **Ampullonatica ambulacrum** (J. Sowerby). (× 1) Becton Formation, Barton, Hampshire. RANGE: Selsey Formation–Becton Formation.

5. **Euspira bassae** Wrigley. (× 1) Thanet Formation, Bishopstone, near Herne Bay, Kent. RANGE: Thanet Formation–Harwich Formation.

6. **Ampullina grossa grossa** (Deshayes). (× 1) Barton Clay Formation, Barton, Hampshire. RANGE: Selsey Formation–Becton Formation. [Syn., *Globularia grossa.*]

7. **Globularia patuloides** (Cossmann & Pissarro). (× 1) Barton Clay Formation, Barton, Hampshire. RANGE: Selsey Formation–Barton Clay Formation. [Syn., *Ampullina patula* (Lamarck) in part, *Globularia patula, Globularia patula brabantica* Glibert.]

8. **Globularia sigaretina** (Lamarck). (× 1) Barton Clay Formation, Barton, Hampshire. RANGE: Earnley Formation–Barton Clay Formation. [Syn., *Ampullina sigaretina.*]

9. **Crommium acutum** (Lamarck). (× 1) Selsey Formation, Bramshaw, Hampshire. RANGE: Earnley Formation–Barton Clay Formation: Elmore Member. [Syn., *Crommium willemeti* (Deshayes), *Ampullina willemeti.*]

10. **Xenophora schroeteri** (Gmelin). (× 1) Selsey Formation, Bramshaw, Hampshire. RANGE: London Clay Formation–Becton Formation, Headon Hill Formation: Colwell Bay Member?. [Syn. *Xenophora agglutinans* (Lamarck).]

Plate 17 continued

14. **Viviparus lentus** (Solander). (× 1) Headon Hill Formation: Totland Bay Member, Headon Hill, Isle of Wight. RANGE: Headon Hill Formation–Bouldnor Formation.

15. **Onustus (Trochotugurium) extensus** (J. Sowerby). (× 1) London Clay Formation, Highgate, Greater London. RANGE: London Clay Formation. [Syn., *Xenophora extensa.*]

Plate 18

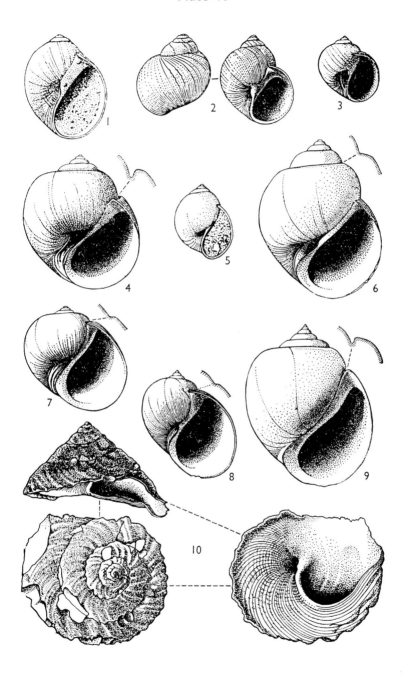

Plate 19
Eocene and Oligocene Gastropods

1. **Cerithiella elongata** (Wrigley). (× 1½) London Clay Formation, Whetstone, Greater London. RANGE: London Clay Formation. [Syn., *Orthochetus elongatus.*]

2. **'Tarebia' acuta** (J. Sowerby). (× 2) Headon Hill Formation, Headon Hill, Isle of Wight. RANGE: Headon Hill Formation–Bouldnor Formation. [Syn., *Melanoides acuta, Melania acuta, Melania muricata* Forbes.]

3, 4. **Potamides (Ptychopotamides) vagus** (Solander). (× 1) Headon Hill Formation: Totland Bay Member, Hordle, Hampshire. RANGE: Headon Hill Formation. [Syn., *Ptychopotamides vagus.*]

5. **Eotympanotonos funatus** (J. Sowerby). (× 1) Woolwich Formation, Newhaven, East Sussex. RANGE: Woolwich Formation–Blackheath Formation. [Syn., *Tympanotonus funatus, Potamides funatus.*]

6. **Granulolabium plicatum** (Bruguière). (× 1½) Bouldnor Formation: Cranmore Member, Hamstead, Isle of Wight. RANGE: Bouldnor Formation. [Syn., *Terebralia plicata, Pirenella monilifera* (Defrance).]

7. **Sigmesalia sp.** (× 1) Selsey Formation, Bracklesham Bay, West Sussex. RANGE: Selsey Formation. [Syn., *Mesalia sulcata* (Lamarck) of authors, *Mesalia incerta* (Deshayes) of authors, *Sigmesalia* aff. *incerta.*]

8. **Melanopsis antidiluviana** (Poiret). (× 1) Blackheath Formation, Abbey Wood, Greater London. RANGE: Woolwich Formation–Blackheath Formation. [Syn., *Melanopsis buccinoidea* (Férussac) of authors.]

9. **Potamaclis turritissima** (Forbes). (× 3) Bouldnor Formation: Hamstead Member, Hamstead, Isle of Wight. RANGE: Headon Hill Formation–Bouldnor Formation.

10. **Vicinocerithium concavum** (J. Sowerby). (× 1) Headon Hill Formation: Colwell Bay Member, Headon Hill, Isle of Wight. RANGE: Headon Hill Formation: Totland Bay Member–Colwell Bay Member. [Syn., *Potamides concavus, Batillaria concava.*]

11. **Haustator sp.** (× 1) Selsey Formation, Bramshaw, Hampshire. RANGE: Selsey Formation. [Syn., *Turritella imbricataria* Lamarck of English authors, in part.]

12. **'Ispharina' sp.** (× ½) Earnley Formation, Bracklesham, West Sussex. RANGE: Earnley Formation–Marsh Farm Formation. [Syn. *Turritella sulcifera* Deshayes of English authors, in part].

13, 14. **'Brotia' melanioides** (J. Sowerby). (× 1) 13, Woolwich Formation, Charlton, Greater London; 14, Epernay Formation, Pourcy, Paris Basin – to show labral profile. RANGE: Woolwich Formation–Blackheath Formation. [Syn., *Melania inquinata* Defrance, *Melanatria inquinata.*]

Plate 19 continued opposite Plate 20

Plate 19

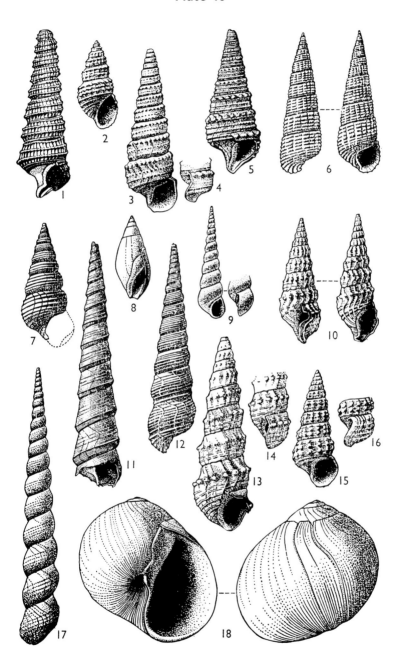

Plate 20
Eocene Gastropods

1. **Aporrhais triangulata** Gardner. (× 1) Harwich Formation: Oldhaven Member, west of Oldhaven Gap, Herne Bay, Kent. RANGE: Blackheath Formation–Harwich Formation.

2. **Rimella rimosa** (Solander). (× 1) Barton Clay Formation, Barton, Hampshire. RANGE: Barton Clay Formation–Headon Hill Formation: Colwell Bay Member.

3, 4. **Eotibia lucida** (J. Sowerby). (× 1½) 3, juvenile shell, 4, adult shell. London Clay Formation, Highgate, Greater London. RANGE: London Clay Formation. [Syn., *Tibia lucida*.]

5, 6. **Aporrhais sowerbii sowerbii** (Fleming). (× 1½) London Clay Formation: 5, Highgate, Greater London; 6, Tolworth, Greater London. RANGE: London Clay Formation.

7. **Dientomochilus bartonense** (J. Sowerby). (× 1½) Barton Clay Formation, Highcliffe, Dorset. RANGE: Barton Clay Formation. [Syn., *Strombus bartonensis*.]

8. **Seraphs sopitum** (Solander). (× 1) Becton Formation?, Barton, Hampshire. RANGE: Selsey Formation–Headon Hill Formation: Colwell Bay Member. [Syn., *Terebellum sopitum*.]

9. **Ectinochilus planum** (Beyrich). (× 1½) Barton Clay Formation: Elmore Member, Huntingbridge, near Fritham, Hampshire. RANGE: Selsey Formation–Barton Clay Formation. [Syn., *Rimella canalis* of authors.]

10. **Hippochrenes amplus** (Solander). (× ¾) Barton Clay Formation, Barton, Hampshire. RANGE: Earnley Formation–Becton Formation.

Plate 19 continued

15, 16. **Varicipotamides ventricosus** (J. Sowerby). (× 2) Headon Hill Formation: Colwell Bay Member, Hordle, Hampshire. RANGE: Headon Hill Formation: Colwell Bay Member. [Syn., *Exechostoma ventricosa, Potamides ventricosus, Batillaria ventricosa*.]

17. **Haustator editus** (Solander). (× 1) Becton Formation, Barton, Hampshire. RANGE: Becton Formation. [Syn. *Turritella edita*.]

18. **Sigatica hantoniensis** (Pilkington). (× 1) London Clay, Portsmouth, Hampshire. RANGE: London Clay Formation–Headon Hill Formation: Colwell Bay Member. [Syn., *Natica hantoniensis*.]

Plate 20

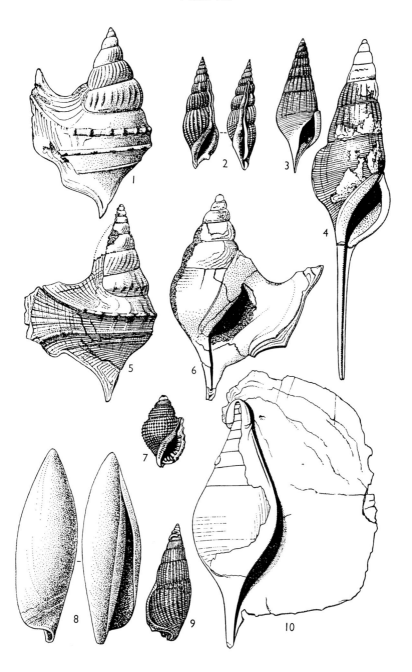

Plate 21
Eocene Gastropods

1. *Ficus nexilis* (Solander). (× 1½) Barton Clay Formation, Barton, Hampshire. RANGE: Selsey Formation–Barton Clay Formation.

2, 3. *Ficopsis multiformis* (Wrigley). (× 1½) London Clay Formation, near Finchley, Greater London. RANGE: London Clay Formation. [Syn., *Ficus multiformis*.]

4. *Galeodea gallica* Wrigley. (× 1½) London Clay Formation, Highgate, Greater London. RANGE: London Clay Formation.

5. *Campanile cornucopiae* (J. Sowerby). (× ½) Selsey Formation, Stubbington, Hampshire. RANGE: Selsey Formation.

6. *Priscoficus smithi* (J. de C. Sowerby). (× ¾) London Clay Formation, Portsmouth Dock, Hampshire. RANGE: London Clay Formation. [Syn., *Ficus smithi*.]

Plate 21

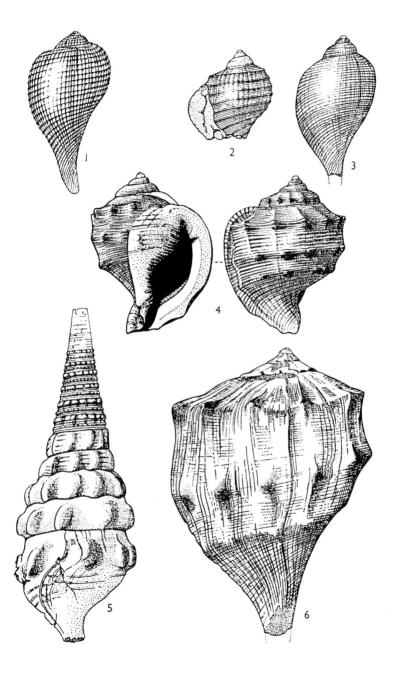

Plate 22
Paleocene and Eocene Gastropods

1. *Pseudoliva fissurata* (Deshayes). (× 1) Blackheath Formation, Abbey Wood, Greater London. RANGE: Blackheath Formation.

2, 3. *Eosiphonalia subnodosa* (Morris). (× 1¼) Harwich Formation: Oldhaven Member; 2, Bishopstone, near Herne Bay, Kent; 3, Swanscombe Hill, Kent. RANGE: Thanet Formation–Harwich Formation. [Syn., '*Siphonalia*' *subnodosa*.]

4. *Sassia arguta arguta* (Solander). (× 1½) Barton Clay Formation, Barton, Hampshire. RANGE: Barton Clay Formation.

5. *Sconsia ambigua* (Solander). (× 1½) Barton Clay Formation, Barton, Hampshire. RANGE: Barton Clay Formation–Headon Hill Formation: Colwell Bay Member. [Syn., *Cassis ambigua*.]

6. *Editharus labiatus* (J. de C. Sowerby). (× 1) Headon Hill Formation: Colwell Bay Member, Brockenhurst, Hampshire. RANGE: Headon Hill Formation. [Syn., *Pollia labiata*.]

7. *Galeodea coronata* (Deshayes). (× 1½) Selsey Formation, Brook, Hampshire. RANGE: Selsey Formation.

Plate 22

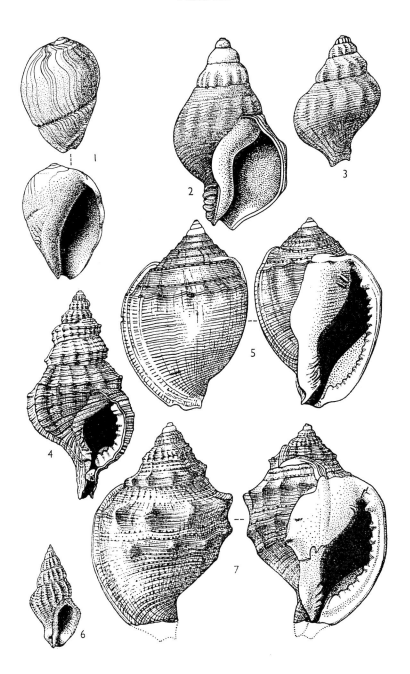

Plate 23
Eocene Gastropods

1. *Eocantharus londini* (Wrigley). (× 1¼) London Clay Formation, Finchley, Greater London. RANGE: London Clay Formation. [Syn., *Pollia londini*, *Tritonidea londini*.]

2. *Timbellus crenulatus crenulatus* (Röding). (× 1½) Barton Clay Formation, Barton, Hampshire. RANGE: Selsey Formation–Becton Formation. [Syn., *Pterynotus tricarinatus* (Lamarck), *Pterynotus tricuspidatus* (Deshayes), *Murex tricarinatus* form *tricuspidatus* Deshayes.]

3. *Typhis pungens* (Solander). (× 1½) Barton Clay Formation, Barton, Hampshire. RANGE: Selsey Formation–Headon Hill Formation: Colwell Bay Member.

4. '*Trophonopsis*' *hantoniensis* (Edwards). (× 1½) Headon Hill Formation: Colwell Bay Member, Brockenhurst, Hampshire. RANGE: Headon Hill Formation: Colwell Bay Member. [Syn., *Murex hantoniensis*, *Pterynotus hantoniensis*.]

5, 6. *Clavilithes pinus* (Perry). (5, initial whorls, enlarged; 6, ×1) Barton Clay Formation, Barton, Hampshire. RANGE: Barton Clay Formation–Becton Formation. [Syn., *Clavilithes macrospira* Cossmann.]

7. *Jsowerbya sexdentata* (J. de C. Sowerby). (× 1½) Headon Hill Formation: Colwell Bay Member, Colwell Bay, Isle of Wight. RANGE: Becton Formation–Headon Hill Formation: Colwell Bay Member. [Syn., '*Murex*' *sexdentatus*, *Urosalpinx sexdentatus*.]

8. *Fusinus asper* (J. Sowerby). (a, × 1½, b, initial whorls of another specimen, enlarged.) Becton Formation, Barton, Hampshire. RANGE: Becton Formation.

9. *Bartonia canaliculata* (J. de C. Sowerby). (× 1½) Barton Clay Formation, Barton, Hampshire. RANGE: Barton Clay Formation. [Syn., *Cominella canaliculata*.]

10. *Clavilithes longaevus* (Solander). (× ½) Barton Clay Formation, Barton, Hampshire. RANGE: Barton Clay Formation.

11. *Fusinus porrectus* (Solander). (× 1) Barton Clay Formation, Barton, Hampshire. RANGE: Barton Clay Formation–Becton Formation.

12. '*Fusinus*' *wetherelli* Wrigley. (× 1) London Clay Formation, Finchley, Greater London. RANGE: London Clay Formation.

Plate 23

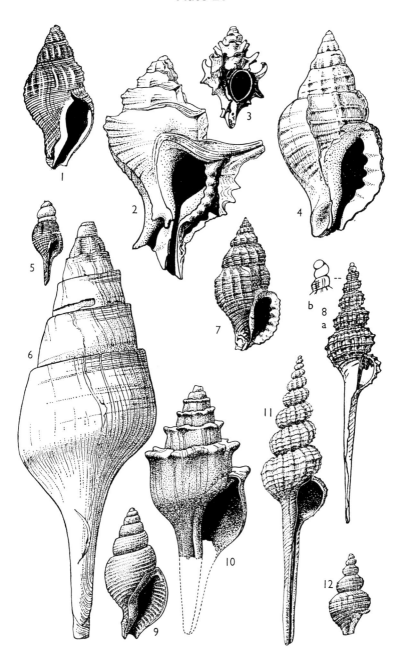

Plate 24
Eocene and Oligocene Gastropods

1. **Streptolathyrus zonulatus** Wrigley. (× 1¾) London Clay Formation, Highgate, Greater London. RANGE: London Clay Formation.

2. **Wrigleya transversaria** (Wrigley). (× 1½) London Clay Formation, Highgate, Greater London. RANGE: London Clay Formation. [Syn., *Euthriofusus transversarius.*]

3. **Streptolathyrus cymatodis** (Edwards). (× 1) London Clay Formation, Clarendon, Wiltshire. RANGE: London Clay Formation.

4. **Wrigleya complanata** (J. de C. Sowerby). (× 1) London Clay Formation, Highgate, Greater London. RANGE: London Clay Formation. [Syn., *Euthriofusus complanatus.*]

5. **Sycostoma pyrus** (Solander). (× 1) Barton Clay Formation, Barton, Hampshire. RANGE: Barton Clay Formation–Becton Formation. [Syn., *Leiostoma pyrus, Sycum pyrus.*]

6. **Bartonia curta** (J. de C. Sowerby). (× 1½) London Clay Formation, Fareham, Hampshire. RANGE: London Clay Formation. [Syn., *Pseudoneptunea curta.*]

7. **Strepsidura turgida** (Solander). (× 1) Selsey Formation, Brook, Hampshire. RANGE: Earnley Formation–Becton Formation.

8. **Volutocorbis lima** (J. de C. Sowerby). (× 1) Barton Clay Formation, Barton, Hampshire. RANGE: Barton Clay Formation–Becton Formation. [Syn., *Volutocorbis scabricula* (Solander *non* Linnaeus).]

9. **Athleta affinis** (Brocchi). (× 1) Bouldnor Formation: Cranmore Member, Hamstead, Isle of Wight. RANGE: Bouldnor Formation: Cranmore Member. [Syn., *Athleta rathieri* Hébert, *Neoathleta rathieri.*]

10. **Surculites errans** (Solander). (× 1½) London Clay Formation, near Chalk Farm, Greater London. RANGE: London Clay Formation–Barton Clay Formation. [Syn., *Surculites bifasciatus* (J. Sowerby) of authors.]

11. **Wrigleya regularis regularis** (J. Sowerby). (× 1) Barton Clay Formation, Barton, Hampshire. RANGE: Barton Clay Formation. [Syn., *Euthriofusus regularis, Chrysodomus antiquus* (Solander *non* Linnaeus).]

12. **Volutospina ambigua** (Solander). (× 1) Barton Clay Formation, Barton, Hampshire. RANGE: Barton Clay Formation–Becton Formation. [Syn., *Volutocorbis ambigua.*]

13. **Cornulina minax** (Solander). (× 1) Barton Clay Formation, Barton, Hampshire. RANGE: London Clay Formation–Headon Hill Formation: Colwell Bay Member.

Plate 24

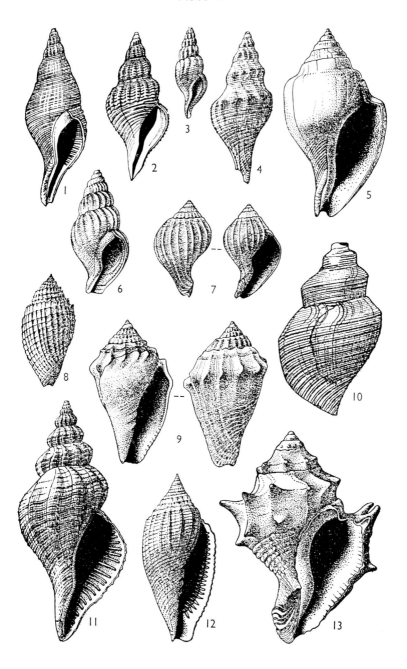

Plate 25
Eocene Gastropods

1. **Volutospina scalaris** (J. de C. Sowerby). (× 1) Becton Formation, Barton, Hampshire. RANGE: Barton Clay Formation–Becton Formation. [Syn., *Athleta scalaris*.]

2. **Volutospina aff. spinosa** (Linnaeus). (× 1) Selsey Formation, Brook, Hampshire. RANGE: Earnley Formation–Barton Clay Formation: Elmore Member. [Syn., *Athleta spinosus*, in part.]

3. **Lyria decora** (Beyrich). (× 1) Headon Hill Formation: Colwell Bay Member, Brockenhurst, Hampshire. RANGE: Selsey Formation–Headon Hill Formation: Colwell Bay Member. [Syn., *Voluta maga* Edwards.]

4. **Volutospina selseiensis** (Edwards). (× 1) Selsey Formation, Brook, Hampshire. RANGE: Marsh Farm Formation–Barton Clay Formation: Elmore Member. [Syn., *Athleta selseiensis, Neoathleta selseiensis.*]

5. **Volutospina denudata** (J. de C. Sowerby). (× 1) London Clay Formation, Bognor Regis, West Sussex. RANGE: London Clay Formation. [Syn., *Athleta denudatus*.]

6. **Conomitra parva** (J. de C. Sowerby). (× 2) Barton Clay Formation, Highcliffe, Dorset. RANGE: Barton Clay Formation–Becton Formation. [Syn., *Mitra parva*.]

7. **Volutospina nodosa** (J. de C. Sowerby). (× 1) London Clay Formation, Highgate, Greater London. RANGE: London Clay Formation–Barton Clay Formation. [Syn., *Athleta nodosus*.]

8–10. **Volutospina luctator** (Solander). (8, × 1; 9, initial whorls × 10; 10, young shell × 1) Barton Clay Formation, Barton, Hampshire. RANGE: Barton Clay Formation–Becton Formation. [Syn., *Athleta luctator*.]

Plate 26
Eocene Gastropods

1. **Olivula canalifera** (Lamarck). (× 1) Becton Formation, Barton, Hampshire. RANGE: Earnley Formation–Becton Formation. [Syn. *Ancilla canalifera, Ancillarina canalifera, Tortoliva canalifera*.]

2. **Conorbis dormitor** (Solander). (× 1) Barton Clay or Becton Formation, Barton, Hampshire. RANGE: Barton Clay Formation–Becton Formation.

Plate 26 continued on next page

Plate 25

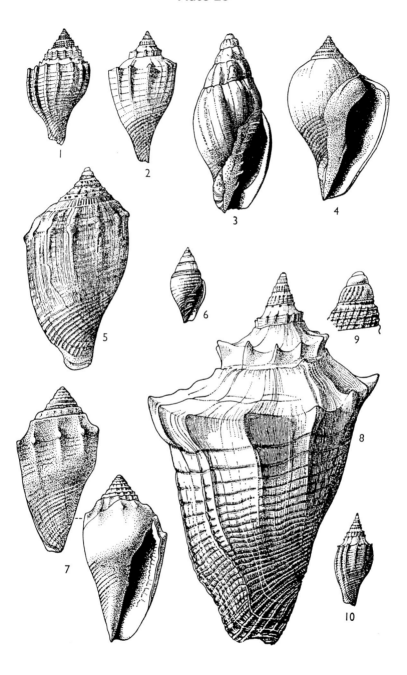

3. ***Pseudolivella branderi*** (J. Sowerby). (× 1) Becton Formation, Barton, Hampshire. RANGE: Barton Clay Formation–Becton Formation. [Syn., *Oliva branderi, Olivella branderi, Callianax branderi, Olivancillaria branderi*.]

4. ***Admetula laeviuscula*** (J. Sowerby). (× 2) London Clay Formation, Highgate, Greater London. RANGE: London Clay Formation. [Syn., *Bonellitia laeviuscula*.]

5. ***Stazzania bifidoplicata*** (Charlesworth). (× 6) Barton Clay Formation, Highcliffe, Dorset. RANGE: Barton Clay Formation–Becton Formation. [Syn., *Marginella bifidoplicata, Volvarinella bifidoplicata*.]

6. ***Sveltella microstoma*** (Charlesworth). (× 2) Barton Clay Formation, Barton, Hampshire. RANGE: Selsey Formation–Becton Formation.

7. ***Admetula evulsa evulsa*** (Solander). (× 1) Barton Clay Formation, Barton, Hampshire. RANGE: Barton Clay Formation–Becton Formation. [Syn., *Bonellitia evulsa*.]

8. ***Mitreola scabra*** (J. de C. Sowerby). (× 1) Becton Formation, Barton, Hampshire. RANGE: Becton Formation. [Syn., *Mitra scabra*.]

9. ***Coptostoma quadratum*** (J. Sowerby). (× 2) Barton Clay Formation, Highcliffe, Dorset. RANGE: Barton Clay Formation.

10. ***Orthosurcula rostrata*** (Solander). (× 1) Barton Clay Formation, Barton, Hampshire. RANGE: Barton Clay Formation. [Syn., *Turricula rostrata*.]

11. ***Orthosurcula teretrium*** (Edwards). (× 1) London Clay Formation, Highgate, Greater London. RANGE: London Clay Formation. [Syn., *Turricula teretrium*.]

12. ***Unitas nassaeformis*** (Wrigley). (× 2½) Becton Formation, Barton, Hampshire. RANGE: Selsey Formation–Becton Formation. [Syn., *Uxia nassaeformis*.]

13. ***Bonellitia pyrgota*** (Edwards). (× 1) Headon Hill Formation: Colwell Bay Member, Colwell Bay, Isle of Wight. RANGE: Headon Hill Formation: Colwell Bay Member.

14, 15. ***Hastula plicatula*** (Lamarck). (× 2) 15 showing remnant of colour banding. Blackheath Formation, Abbey Wood, Greater London. RANGE: Blackheath Formation. [Syn., *Terebra plicatula, Terebrellina plicatula*.]

16, 17. ***Bathytoma turbida*** (Solander). (× 1) Barton Clay Formation, Barton, Hampshire. RANGE: Barton Clay Formation–Becton Formation.

18. ***Volutospina athleta*** (Solander). (× ¾) Barton Clay Formation, Barton, Hampshire. RANGE: Barton Clay Formation. [Syn., *Athleta athleta*.]

19. ***Eopleurotoma simillima*** (Edwards). (× 1½) London Clay Formation, Clarendon, Wiltshire. RANGE: London Clay Formation. [Syn., *Fusiturris simillima*.]

Plate 26

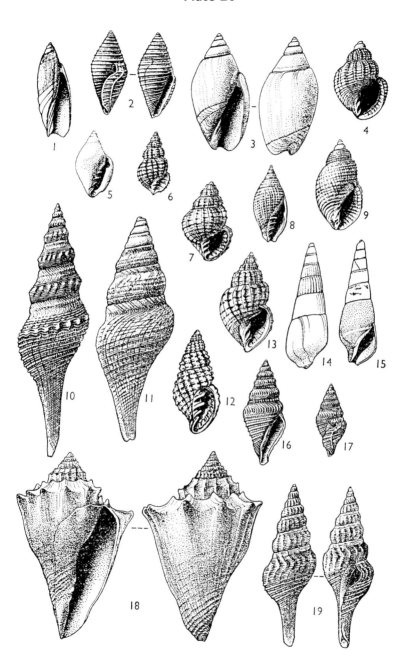

Plate 27
Eocene and Oligocene Gastropods

1. **Palaeoxestina occlusa** (Edwards). (× 1) Bembridge Limestone Formation, Headon Hill, Isle of Wight. RANGE: Bembridge Limestone Formation. [Syn., *Helix occlusa*.]

2. **Megalocochlea pseudoglobosa** (d'Orbigny). (× ½) Bembridge Limestone Formation, Tapnell, near Freshwater, Isle of Wight. RANGE: Bembridge Limestone Formation. [Syn., *Helix pseudoglobosa*.]

3. **Klikia vectiensis** (Edwards). (× 1½) Bembridge Limestone Formation, Headon Hill, Isle of Wight. RANGE: Bembridge Limestone Formation. [Syn., *Helix vectiensis*.]

4. **Planorbarius discus** (Edwards). (× 1) Bembridge Limestone Formation, Sconce, Isle of Wight. RANGE: Headon Hill Formation–Bouldnor Formation: Hamstead Member. [Syn., *Planorbis discus*.]

5. **Tornatellaea simulata** (Solander). (× 1½) Becton Formation, Barton, Hampshire. RANGE: London Clay Formation–Headon Hill Formation: Colwell Bay Member.

6. **'Gemmula' plebeia** (J. de C. Sowerby). (× 1½) Selsey Formation?, Bracklesham Bay, West Sussex. RANGE: Earnley Formation–Barton Clay Formation: Elmore Member. [Syn., *Pleurotoma plebeia*, *Pleurotoma denticula* of authors, *Hemipleurotoma plebeia*.]

7. **Lymnaea (Galba) longiscata** (Brongniart). (× 1) Headon Hill Formation, Headon Hill, Isle of Wight. RANGE: Headon Hill Formation–Bouldnor Formation. [Syn., *Lymnaea longiscata*.]

8. **Australorbis euomphalus** (J. Sowerby). (× 1) Headon Hill Formation, Headon Hill, Isle of Wight. RANGE: Headon Hill Formation. [Syn., *Planorbis euomphalus*.]

9. **'Conus' edwardsi** (Cossmann). (× ¾) Selsey Formation, Bramshaw, Hampshire. RANGE: Selsey Formation. [Syn., *Leptoconus edwardsi*, *Conus deperditus* Bruguière of Edwards.]

10. **Hemiconus scabriculus** (Solander). (× 1½) Becton Formation, Barton, Hampshire. RANGE: Barton Clay Formation–Becton Formation.

11. **Rillyarex ellipticus** (J. Sowerby). (× ½) Bembridge Limestone Formation, Headon Hill, Isle of Wight. RANGE: Headon Hill Formation–Bembridge Limestone Formation. [Syn., *'Filholia' elliptica*, *Bulimus ellipticus*.]

12. **Palaeoglandina costellata** (J. Sowerby). (× 1) Bembridge Limestone Formation, Sconce, Isle of Wight. RANGE: Bembridge Limestone Formation. [Syn., *Bulimus costellatus*.]

Plate 27

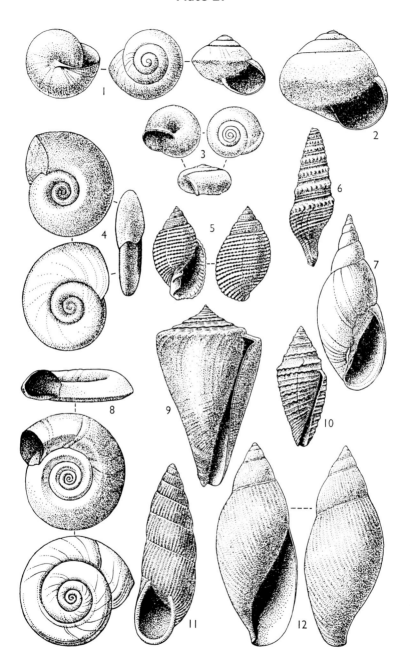

Plate 28
Paleocene and Eocene Sharks' Teeth (Figs 1–6, 8–10), Cephalopods (Figs 7, 12) and Scaphopod (Fig. 11)

1. **Otodus obliquus** Agassiz. (× 1) London Clay Formation, Isle of Sheppey, Kent. RANGE: Thanet Formation–London Clay Formation. [Syn., *Lamna obliqua*.]

2. **Sylvestrilamia teretidens** (White). (× 1½) Blackheath Formation, Abbey Wood, Greater London. RANGE: Thanet Formation–Harwich Formation. [Syn., *Odontaspis teretidens*.]

3. **Striatolamia macrota** (Agassiz) (× 1½) Blackheath Formation, Abbey Wood, Greater London. RANGE: Blackheath Formation–Barton Clay Formation. [Syn., *Odontaspis elegans* (Agassiz) of authors.]

4, 5. **Physogaleus secundus** (Winkler). (× 2) 4, Barton Clay Formation, Barton, Hampshire; 5, Earnley Formation, Southampton, Hampshire. RANGE: London Clay Formation–Barton Clay Formation. [Syn., *Physodon secundus, Galeocerdo minor, Galeorhinus minor, Eugaleus minor*.]

6. **Galeocerdo latidens** Agassiz. (× 1½) Earnley Formation, Southampton, Hampshire. RANGE: Earnley Formation–Barton Clay Formation.

7. **Belosaepia sepioidea** (Blainville). (× 1) Selsey Formation, Bramshaw, Hampshire. RANGE: London Clay Formation–Barton Clay Formation. [Syn., *Belosepia sepioidea*.]

8. **Otodus auriculatus** (Blainville). (× 1) Earnley Formation, Bracklesham, West Sussex. RANGE: Earnley Formation–Barton Clay Formation. [Syn., *Carcharocles auriculatus, Procarcharodon auriculatus, Carcharodon auriculatus*.]

9. **Notorynchus serratissimus** (Agassiz). (× 1½) London Clay Formation, Isle of Sheppey, Kent. RANGE: London Clay Formation. [Syn., *Notidanus serratissimus*.]

10. **Squatina prima** (Winkler). (× 1½) Blackheath Formation, Abbey Wood, Greater London. RANGE: Thanet Formation–Becton Formation.

11. **Antalis bartonensis** (Palmer). (× 1) Barton Clay Formation, Barton, Hampshire. RANGE: Barton Clay Formation–Becton Formation. [Syn., *Dentalium striatum* J. Sowerby non Born, *Antalis striata, Dentalium bartonense*.]

12. **Cimomia imperalis** (J. Sowerby). (× ¾) Shell-wall broken away to show the internal septa. London Clay Formation, Isle of Sheppey, Kent. RANGE: London Clay Formation–Selsey Formation? [Syn., *Nautilus imperialis*.]

Plate 28

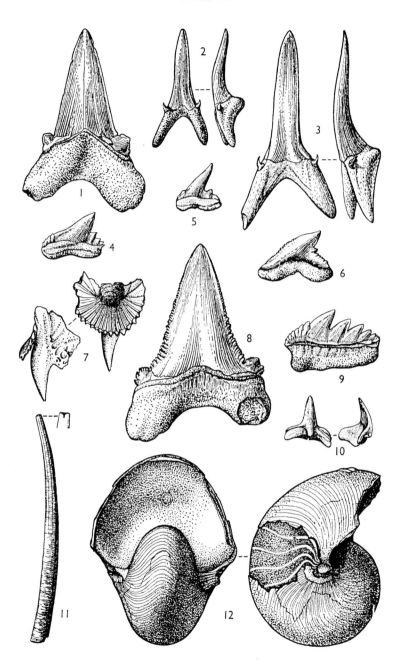

Plate 29
Eocene and Oligocene Fishes

1. ***Aetobatus irregularis*** Agassiz. Eagle Ray. Lower tooth-plate. (× ½) a, anterior edge; b, grinding surface; c, attached surface. Earnley Formation, Southampton, Hampshire. RANGE: London Clay Formation–Becton Formation.

2. ***Myliobatis striatus*** Buckland. Eagle Ray. Lower tooth-plate. (× ½) a, anterior edge; b, attached surface; c, grinding surface. Barton Clay Formation, Highcliffe, Dorset. RANGE: Earnley Formation–Becton Formation.

3. ***Hypolophodon sylvestris*** (White). Tooth. (× 2) Blackheath Formation, Abbey Wood, Greater London. RANGE: Upnor Formation–Harwich Formation.

4. ***Acipenser* sp.** Sturgeon. Lateral scute. (× 1) Bouldnor Formation: Hamstead Member, Hamstead, Isle of Wight. RANGE: of genus: Late Cretaceous–present day.

Plate 29

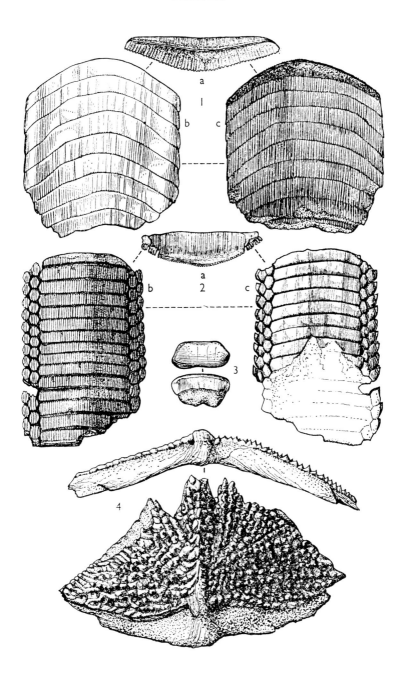

Plate 30
Paleocene, Eocene and Oligocene Fishes (Figs 1–6), Reptiles (Figs 7–10) and Mammal (Fig. 11)

1. **Phyllodus toliapicus** Agassiz. Upper pharyngeal dentition. (× 1) Blackheath Formation, Abbey Wood, Greater London. RANGE: Upnor Formation–London Clay Formation.

2. **Edaphodon bucklandi** Agassiz. Mandibular tooth. (× ½) Earnley Formation, Bracklesham Bay, West Sussex. RANGE: Upnor Formation–Selsey Formation.

3. **Amia sp.** Opercular plate. (× 1) Bouldnor Formation: Bembridge Marls Member, Thorness Bay, Isle of Wight. RANGE: of genus: Late Cretaceous–present day (freshwater).

4. **Albula eppsi** White & Frost. Otolith. (× 1) Blackheath Formation, Abbey Wood, Greater London, Kent. RANGE: Blackheath Formation–Harwich Formation.

5. **Lepisosteus suessionensis** Gervais. Scale. (× 1) Blackheath Formation, Abbey Wood, Greater London. RANGE: Upnor Formation–Harwich Formation. [Syn., *Lepidosteus suessionensis*.]

6. **Cylindracanthus rectus** (Dixon). Fish rostrum. (× ½) Earnley Formation, Bracklesham, West Sussex. RANGE: London Clay Formation–Becton Formation.

7. **Palaeophis toliapicus** Owen. Vertebra of Snake. (× 1) London Clay Formation, Isle of Sheppey, Kent. RANGE: London Clay Formation.

8, 9. **Diplocynodon hantoniensis** (Wood). Crocodile. (8, tooth × 1; 9, dorsal vertebra × ½). Headon Hill Formation: Totland Bay Member, Hordle, Hampshire. RANGE: Headon Hill Formation–Bouldnor Formation.

10. **Rafetoides henrici** (Owen). Costal bone of Turtle. (× ½) Headon Hill Formation: Totland Bay Member, Hordle, Hampshire. RANGE: Headon Hill Formation: Totland Bay Member. [Syn., *Trionyx circumsulcatus* Owen, *Aulacochelys circumsulcatus*.]

11. **Plagiolophus minor** (Cuvier). Upper molar of early relative of Horse. (× 1) Bembridge Limestone Formation, Headon Hill, Isle of Wight. RANGE: Bembridge Limestone Formation–Bouldnor Formation: Hamstead Member. [Syn., *Palaeotherium minus*.]

Plate 30

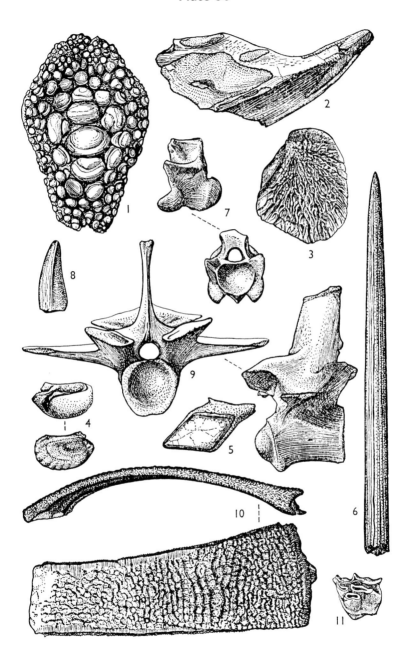

Plate 31

Pliocene and Quaternary Bryozoan (Fig. 1), Corals (Figs 2, 4) and Echinoderm (Fig. 3)

1. **Meandropora tubipora** (Busk). (a, × 1; b, × 5). Coralline Crag Formation, Sudbourne, Suffolk. RANGE: Coralline Crag Formation. [Syn., *Fascicularia tubipora*.]

2. **Sphenotrochus intermedius** (Münster). (× 3) Coralline Crag Formation, Sutton, Suffolk. RANGE: Coralline Crag Formation–Red Crag Formation.

3. **Echinocyamus pusillus** (Müller). (× 3) Red Crag Formation, Alderton, Suffolk. RANGE: Red Crag Formation–present day.

4. **Balanophyllia calyculus** Wood. (× 3) Red Crag Formation, Bentley, Suffolk. RANGE: Red Crag Formation. [Syn., *Balanophyllia caliculus*.]

Plate 31

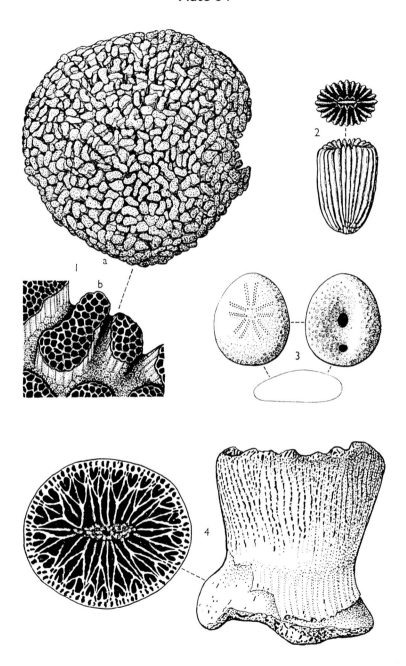

Plate 32

Pliocene and Quaternary Bivalve (Fig. 1), Barnacle (Fig. 2), Brachiopod (Fig. 3) and Echinoderm (Fig. 4)

1. ***Yoldia oblongoides*** (Wood). (× 1) Norwich Crag Formation, Chillesford, Suffolk. RANGE: Norwich Crag Formation–Corton Formation. [Syn., *Yoldia myalis* (Couthouy) of British authors.]

2. ***Concavus concavus*** (Bronn). (× 1) Coralline Crag Formation, Sudbourne, Suffolk. RANGE: Coralline Crag Formation–Pleistocene. [Syn., *Balanus concavus*.]

3. ***Terebratula maxima*** Charlesworth. (× ½) Coralline Crag Formation, Orford, Suffolk. RANGE: Coralline Crag Formation.

4. ***Temnechinus excavatus*** Forbes. (a, b, × 1½; c, × 5). Coralline Crag Formation, Sutton, Suffolk. RANGE: Coralline Crag Formation. [Syn., *Temnopleurus woodii* Agassiz.]

Plate 32

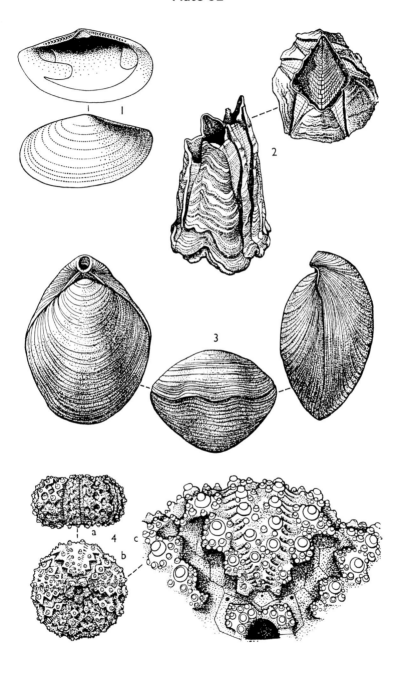

Plate 33
Pliocene and Quaternary Bivalves

1, 2. **Leionucula laevigata** (J. Sowerby). (× 1) Red Crag Formation, Walton-on-the-Naze, Essex. RANGE: Red Crag Formation. [Syn., *Nucula laevigata*.]

3. **Acila cobboldiae** (J. Sowerby). (× 1) Red Crag Formation, Sutton, Suffolk. RANGE: Red Crag Formation–Corton Formation. [Syn., *Nucula cobboldiae*.]

4, 5. **Palliolum tigerinum** (Müller). (4a, × 1½; 4b, ×4; 5, × 1½) Coralline Crag Formation, Ramsholt, Suffolk. RANGE: Coralline Crag Formation–present day. [Syn., *Chlamys tigerina, Eburneopecten tigerinus*.]

6. **Glycymeris variabilis** (J. de C. Sowerby). (× ¾) Red Crag Formation, Walton-on-the-Naze, Essex. RANGE: Coralline Crag Formation–Red Crag Formation. [Syn., *Glycymeris glycymeris* (Linnaeus) in part.]

7, 8. **Aequipecten opercularis** (Linnaeus). (× ¾) Coralline Crag Formation, Sutton, Suffolk. RANGE: Coralline Crag Formation–present day. [Syn., *Pecten opercularis, Chlamys opercularis*.]

Plate 33

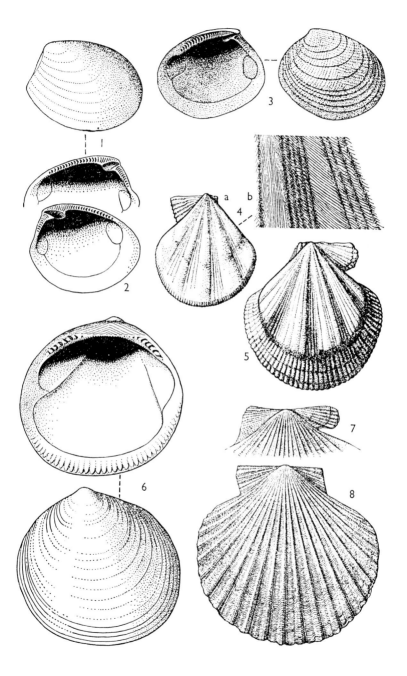

Plate 34
Pliocene and Quaternary Bivalves

1, 2. ***Palliolum gerardi*** (Nyst). (× ¾) Coralline Crag Formation, Gedgrave, Suffolk. RANGE: Coralline Crag Formation. [Syn., *Pseudamussium gerardi*.]

3, 4. ***Digitariopsis obliquata obliquata*** (J. Sowerby). (× 1) Red Crag Formation, Walton-on-the-Naze, Essex. RANGE: Red Crag Formation. [Syn., *Astarte obliquata*.]

5. ***Tridonta borealis*** (Schumacher). (× 1) Pleistocene, Bridlington, Yorkshire. RANGE (in Britain): Norwich Crag Formation–Clyde Valley Formation; Arctic at present day. [Syn., *Astarte borealis, Astarte semisulcata* (Leach).]

6. ***Mytilus* cf. *trossulus*** Gould. (× ¾) Red Crag Formation, Sutton, Suffolk. RANGE: Red Crag Formation–present day. [Syn. *Mytilus edulis* Linnaeus of authors.]

7, 8. ***Ostrea edulis*** Linnaeus. (× ½) Coralline Crag Formation, Suffolk. RANGE: Coralline Crag Formation–present day.

Plate 34

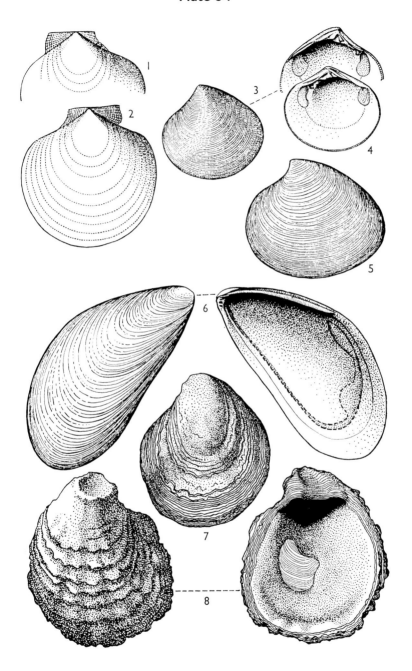

Plate 35
Pliocene and Quaternary Bivalves

1. ***Centrocardita squamulosa ampla*** (Chavan & Coatman). (× 1) Coralline Crag Formation, Orford, Suffolk. RANGE: Coralline Crag Formation–Red Crag Formation. [Syn., *Cardites squamulosa ampla, Centrocardita senilis* (Lamarck) of authors.]

2. ***Cyclocardia scalaris*** (J. Sowerby). (× 2) Red Crag Formation, Ramsholt, Suffolk. RANGE: Coralline Crag Formation–Corton Formation.

3. ***Laevastarte omalii*** (De la Jonkaire). (× 1) Red Crag Formation, Sutton, Suffolk. RANGE: Coralline Crag Formation–Corton Formation. [Syn., *Astarte omalii.*]

4, 5. ***Arctica islandica*** (Linnaeus). (× ¾) Coralline Crag Formation, Ramsholt, Suffolk. RANGE: Coralline Crag Formation–present day.

6. ***Laevastarte mutabilis mutabilis*** (Wood). (× 1) Coralline Crag Formation, Sudbourne, Suffolk. RANGE: Coralline Crag Formation–Red Crag Formation. [Syn., *Astarte mutabilis.*]

7. ***Digitaria digitaria*** (Linnaeus). (× 2) Red Crag Formation, Little Oakley, Essex. RANGE (in Britain): Coralline Crag Formation–Corton Formation; Mediterranean at present day. [Syn., *Astarte digitaria, Woodia digitaria.*]

8. ***Pteromeris corbis*** (Philippi). (× 2) Coralline Crag Formation, Sutton, Suffolk. RANGE (in Britain): Coralline Crag–Corton Formation; Mediterranean at present day. [Syn., *Cardita corbis.*]

Plate 35

Plate 36
Pliocene and Quaternary Bivalves

1, 2. ***Cerastoderma parkinsoni*** (J. Sowerby). (× ¾) Red Crag Formation, Walton-on-the-Naze, Essex. RANGE: Red Crag Formation.

3. ***Cerastoderma hostei*** Chavan. (× ¾) Red Crag Formation, Sutton, Suffolk. RANGE: Coralline Crag Formation–Red Crag Formation. [Syn., *Cerastoderma edule hostei, Cerastoderma edule* (Linnaeus) of authors.]

4. ***Cerastoderma angustatum*** (J. Sowerby). (× 1) Red Crag Formation, Waldringfield, Suffolk. RANGE: Red Crag Formation–Norwich Crag Formation.

5–7. ***Lucinoma borealis*** (Linnaeus). (× 1) Coralline Crag Formation, Ramsholt, Suffolk. RANGE: Coralline Crag Formation–present day. [Syn., *Lucina borealis, Phacoides borealis.*]

8, 9. ***Venus casina casina*** Linnaeus. (× 1) Red Crag Formation, Sutton, Suffolk. RANGE: Coralline Crag Formation–present day.

Plate 36

Plate 37
Pliocene and Quaternary Bivalves

1–3. **Dosinia exoleta** (Linnaeus). (× 1) Red Crag Formation, Walton-on-the-Naze, Essex. RANGE: Red Crag Formation–present day. [Syn., *Artemis exoleta*.]

4. **Scrobicularia plana** (Da Costa). (× 1) Late Middle Pleistocene or Holocene, Selsey Bill, West Sussex. RANGE: Red Crag Formation–present day. [Syn., *Scrobicularia piperata* (Gmelin).]

5. **Spisula arcuata** (J. Sowerby). (× ¾) Red Crag Formation, Walton-on-the-Naze, Essex. RANGE: Coralline Crag Formation–Norwich Crag Formation. [Syn., *Mactra arcuata*.]

Plate 37

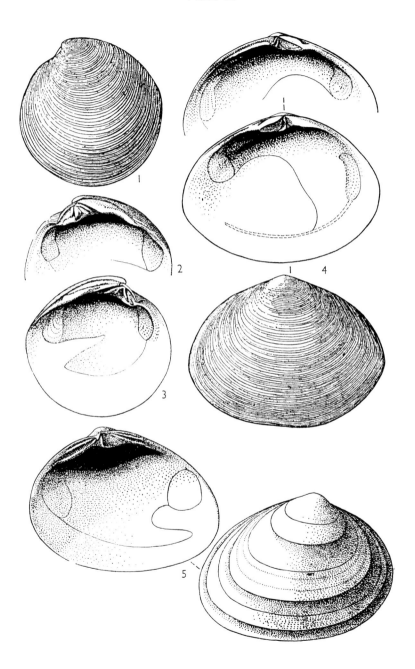

Plate 38
Pliocene and Quaternary Bivalves

1. **Macoma praetenuis** (Woodward). (× 1) Red Crag Formation, Sutton, Suffolk. RANGE: Red Crag Formation–Norwich Crag Formation.

2. **Macoma calcarea** (Gmelin). (× 1) Norwich Crag Formation, Chillesford, Suffolk. RANGE (in Britain): Red Crag Formation–Clyde Valley Formation; Arctic at present day.

3. **Spisula subtruncata** (Da Costa). (× 1) Norwich Crag Formation, Yarn Hill, near Southwold, Suffolk. RANGE: Norwich Crag Formation–present day. [Syn., *Mactra subtruncata*.]

4–6. **Macoma balthica** (Linnaeus). (× 1) 'Shelly drift', Gloppa, near Oswestry, Shropshire. RANGE: Wroxham Crag Formation–present day.

7–9. **Macoma obliqua** (J. Sowerby). (7, × ¾; 8, 9, × 2). Red Crag Formation, Sutton, Suffolk. RANGE: Coralline Crag Formation–Corton Formation.

10, 11. **Mya truncata** Linnaeus. (× ¾) Coralline Crag Formation, Ramsholt, Suffolk. RANGE: Coralline Crag Formation–present day.

Plate 38

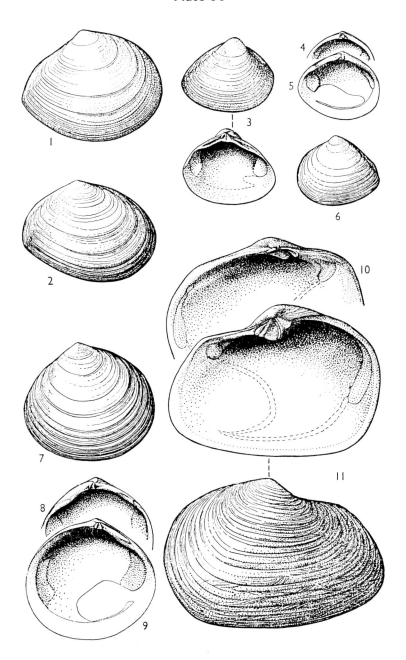

Plate 39
Pliocene and Quaternary Gastropods

1. ***Calliostoma subexcavatum*** Wood. (× 1) Red Crag Formation, Sutton, Suffolk. RANGE: Coralline Crag Formation–Red Crag Formation.

2. ***Emarginula fissura*** (Linnaeus). (× 1½) Red Crag Formation, East Anglia. RANGE: Coralline Crag Formation–present day. [Syn., *Emarginula reticulata* J. Sowerby.]

3. ***Boreoscala greenlandica*** (Perry). (× 1) Norwich Crag Formation, Bramerton, Norfolk. RANGE (in Britain): Norwich Crag Formation– Clyde Valley Formation; Arctic at present day. [Syn., *Epitonium greenlandicum*.]

4. ***Nassarius granulatus*** (J. Sowerby). (× 3) Red Crag Formation, Walton-on-the-Naze, Essex. RANGE: Coralline Crag Formation–Corton Formation. [Syn., *Hinia granulata*.]

5. '***Turritella*** ' *incrassata* J. Sowerby. (× 1) Coralline Crag Formation, Ramsholt, Suffolk. RANGE: Coralline Crag Formation–Corton Formation.

6. '***Turritella*** ' *communis* Risso. (× 1) Glacial deposits, Worden Hall, Lancashire. RANGE: Corton Formation–present day.

7, 8. ***Potamides (Ptychopotamides) tricinctus*** (Brocchi). (× 1) 7, Red Crag Formation, Sutton, Suffolk; 8, Norwich Crag Formation, Yarn Hill, near Southwold, Suffolk. RANGE: Coralline Crag Formation–Corton Formation. [Syn., *Ptychopotamides tricinctus*.]

9. ***Calyptraea chinensis*** (Linnaeus). (× 1) Coralline Crag Formation, Gomer, near Gedgrave, Suffolk. RANGE: Coralline Crag Formation–present day.

10. ***Trivia coccinelloides coccinelloides*** (J. Sowerby). (× 1½) Red Crag Formation, Little Oakley, Essex. RANGE: Coralline Crag Formation–Red Crag Formation. [Syn., *Trivia europaea* Montagu of authors.]

11. ***Euspira hemiclausa*** (J. Sowerby). (× 1) Red Crag Formation, Walton-on-the-Naze, Essex. RANGE: Red Crag Formation–Wroxham Crag Formation. [Syn., *Natica hemiclausa, Polinices hemiclausus*.]

12. ***Capulus ungaricus*** (Linnaeus). (× ¾) Red Crag Formation, Walton-on-the-Naze, Essex. RANGE: Coralline Crag Formation–present day.

13. ***Natica crassa*** Nyst. (× 1) Red Crag Formation, Walton-on-the-Naze, Essex. RANGE: Coralline Crag Formation–Red Crag Formation. [Syn., *Natica multipunctata* Wood *non* de Blainville.]

Plate 39 continued opposite Plate 40

Plate 39

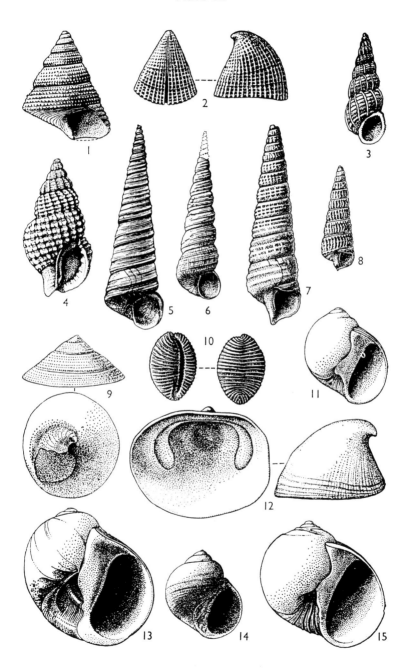

Plate 40
Pliocene and Quaternary Gastropods

1. *Nassarius reticosus* (J. Sowerby). (× 1) Red Crag Formation, Walton-on-the-Naze, Essex. RANGE: Coralline Crag Formation–Corton Formation. [Syn., *Hinia reticosa*, *Uzita reticosa*.]

2. *Buccinum undatum* Linnaeus. (× ¾) Holocene clay, Belfast. RANGE: Red Crag Formation–present day.

3. *Boreotrophon clathratus* (Linnaeus). (× 1) Holderness Formation ('Bridlington Crag'), Bridlington, Yorkshire. RANGE (in Britain): Coralline Crag Formation–Corton Formation; Arctic at present day. [Syn., *Trophonopsis clathratus*, *Trophon scalariforme* (Gould).]

4. *Liomesus dalei* (J. Sowerby). (× 1) Red Crag Formation, Walton-on-the-Naze, Essex. RANGE: Coralline Crag Formation–Red Crag Formation. [Syn., *Buccinum dalei*, *Leiomesus* (sic) *dalei*.]

5. *Neptunea lyratodespecta lyratodespecta* Strauch. (× ¾) Red Crag Formation, Little Oakley, Essex. RANGE: Red Crag Formation. [Syn., *Neptunea despecta decemcostata* (Say) of authors.]

6. *Searlesia costifera* (Wood). (× 1) Red Crag Formation, Walton-on-the-Naze, Essex. RANGE (in Britain): Coralline Crag Formation–Red Crag Formation; North Atlantic at present day. [Syn., *Trophon costiferum*.]

7. *Neptunea angulata* Harmer. (× ½) Red Crag Formation, Walton-on-the-Naze, Essex. RANGE: Red Crag Formation. [Syn., *Neptunea contraria* (Linnaeus) of authors.]

8. *Spinucella tetragona* (J. Sowerby). (× 1) Red Crag Formation, Walton-on-the-Naze, Essex. RANGE: Coralline Crag Formation–Red Crag Formation. [Syn., *Nucella tetragona*.]

9. *Colus curtus* (Jeffreys). (× ¾) Red Crag Formation, Ramsholt, Suffolk. RANGE: Red Crag Formation. [Syn., *Sipho curtus*.]

Plate 39 continued

14. *Littorina littorea* (Linnaeus). (× 1) Red Crag Formation, Sutton, Suffolk. RANGE: Red Crag Formation–present day.

15. *Euspira catenoides* (Wood). (× 1) Red Crag Formation, Walton-on-the-Naze, Essex. RANGE: Coralline Crag Formation–Red Crag Formation. [Syn., *Lunatia catenoides*.]

Plate 40

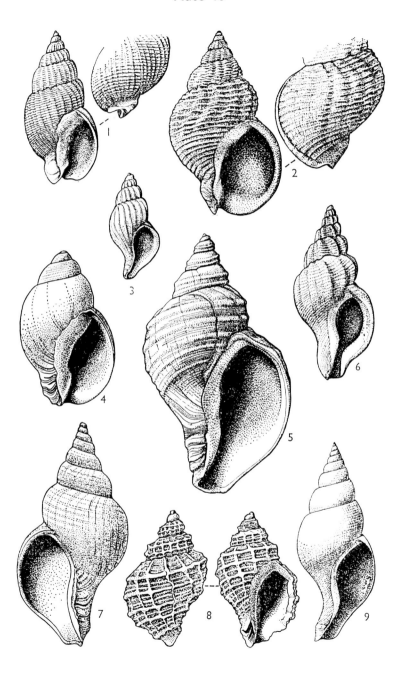

Plate 41

Pliocene and Quaternary Marine Gastropods (Figs 1, 5, 8, 14), Freshwater Gastropods (Figs 6, 7, 9–13, 15), Marine Fishes (Figs 2, 3) and Marine Mammal (Fig. 4)

1. *Nucella incrassata* (J. Sowerby). (× 1) Red Crag Formation, Sutton, Suffolk. RANGE: Red Crag Formation–Corton Formation. [Syn., *Nucella lapillus* (Linnaeus) of authors in part.]

2. *Raja clavata* Linnaeus. Dermal tubercle of Thornback Ray. (× 1½) Norwich Crag Formation, Norwich, Norfolk. RANGE: Coralline Crag Formation–present day.

3. *Carcharodon hastalis* (Agassiz). Shark's tooth. (× 1) Red Crag Formation, Woodbridge, Suffolk. RANGE: Coralline Crag Formation–Red Crag Formation. [Syn., *Isurus hastalis*.]

4. *Balaena affinis* Owen. Ear-bone of Whale. (× ½) Red Crag Formation, Woodbridge, Suffolk. RANGE: Red Crag Formation.

5. *Euroscaphella lamberti* (J. Sowerby). (× 1) Red Crag Formation, Walton-on-the-Naze, Essex. RANGE: Coralline Crag Formation–Red Crag Formation. [Syn., *Scaphella lamberti*.]

6. *Ancylus fluviatilis* Müller. (× 2) Pleistocene (Hoxnian), Clacton, Essex. RANGE: Cromer Forest-bed Formation–present day.

7. *Belgrandia marginata* (Michaud). (× 9) Pleistocene (Hoxnian), Clacton, Essex. RANGE (in Britain): Cromer Forest-bed Formation–Late Pleistocene; southern France at present day.

8. *Admete viridula* (Fabricius). (× 1) Red Crag Formation, Suffolk. RANGE (in Britain): Coralline Crag Formation–Wroxham Crag Formation; Arctic at present day. [Syn., *Cancellaria viridula*, *Admete abnormis* Harmer.]

9. *Radix balthica* (Linnaeus). (× 1) Late Middle Pleistocene, Ilford, Essex. RANGE: Red Crag Formation–present day. [Syn. *Lymnaea peregra* (Müller).]

10. *Valvata piscinalis* (Müller) ('antiqua' form).(× 2½) Pleistocene (Hoxnian), Swanscombe, Kent. RANGE: Middle Pleistocene. [Syn. *Valvata antiqua* Morris.]

11, 12. *Bithynia tentaculata* (Linnaeus). Shell and operculum. (× 2) Pleistocene (Hoxnian), Swanscombe, Kent. RANGE: Norwich Crag Formation–present day.

13. *Viviparus diluvianus* (Kunth). (× 1) Pleistocene (Hoxnian), Swanscombe, Kent. RANGE: Middle Pleistocene. [Syn., *Viviparus clactonensis* (Wood).]

Plate 41 continued opposite Plate 42

Plate 41

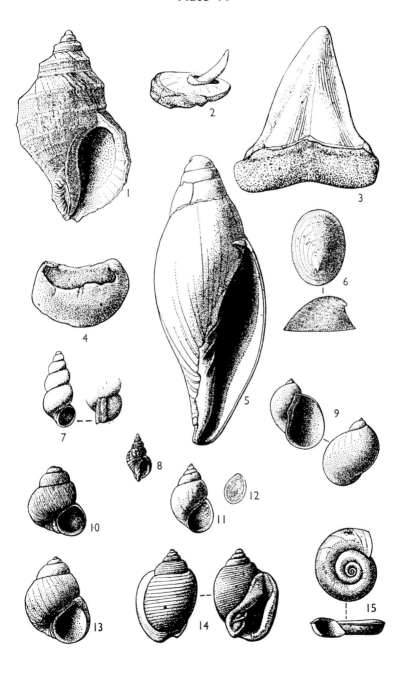

Plate 42
Quaternary Land Gastropods (Figs 1–4) and Freshwater Bivalves (Figs 5–9)

1. **Succinella oblonga** (Draparnaud). (× 3) Pleistocene (Devensian), Tottenham, Greater London. RANGE: Norwich Crag Formation–present day. [Syn., *Succinea oblonga*.]

2. **Trochulus hispidus** (Linnaeus). (× 3) Holocene, Halling, Kent. RANGE: Red Crag Formation–present day. [Syn., *Hygromia hispida, Trichia hispida*.]

3. **Pupilla muscorum** (Linnaeus). (× 10) Pleistocene (Devensian), Ponders End, Greater London. RANGE: Red Crag Formation–present day. [Syn., *Pupa muscorum, Pupa marginata* Draparnaud.]

4. **Cepaea nemoralis** (Linnaeus). (× 1) Late Middle Pleistocene, Ilford, Greater London. RANGE: Red Crag Formation–present day. [Syn., *Helix nemoralis*.]

5, 6. **Potomida littoralis littoralis** (Cuvier). (× ¾) Pleistocene (Hoxnian), Clacton, Essex. RANGE (in Britain): Middle–Late Pleistocene; European continent at present day. [Syn., *Unio littoralis*.]

7. **Pisidium clessini** Neumayr. (× 4) Pleistocene (Hoxnian), Swanscombe, Kent. RANGE: Norwich Crag–late Middle Pleistocene. [Syn., *Pisidium astartoides* Sandberger of authors.]

8, 9. **Corbicula fluminalis** (Müller) ? (× 1) Late Middle Pleistocene, Crayford, Kent. RANGE (in Britain): Norwich Crag Formation–late Middle Pleistocene; North Africa and Middle East at present day.

Plate 41 continued

14. **Ringicula (Ringiculina) ventricosa** (J. de C. Sowerby). (× 3) Red Crag Formation, Sutton, Suffolk. RANGE: Coralline Crag Formation–Norwich Crag Formation.

15. **Planorbis planorbis** (Linnaeus). (× 1½) Pleistocene, West Wittering, West Sussex. RANGE: Red Crag Formation–present day.

Plate 42

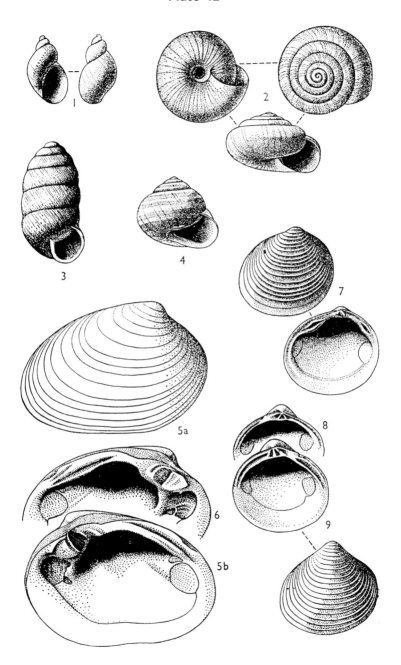

Plate 43
Quaternary Mammals

1, 2. **Equus sp**. Horse. 1. Upper cheek tooth, a, side view and b, biting surface. (× ½) Pleistocene, Felixstowe, Suffolk. 2. Lower cheek tooth, biting surface. (× ½) Pleistocene, dredged off Happisburgh, Norfolk. RANGE: Late Pliocene–present day.

3, 4. **Bos primigenius** Bojanus. Aurochs or Wild Ox. 3. Upper molar, biting surface. (× ½) Pleistocene (Devensian) or Holocene, Walthamstow, Greater London. 4. Third lower molar, a, biting surface and b, side view. (× ½) Pleistocene, Kent's Cavern, Torquay, Devon. RANGE (in Britain): late Middle Pleistocene–Bronze Age; domesticated during Neolithic.

5. **Coelodonta antiquitatis** (Blumenbach). Upper molar of Woolly Rhinoceros. (× ½) Pleistocene (Devensian), Kent's Cavern, Torquay, Devon. RANGE: Middle-Late Pleistocene. [Syn., *Rhinoceros antiquitatis, Tichorhinus antiquitatis, Rhinoceros tichorhinus* Cuvier.]

6. **Hippopotamus amphibius** Linnaeus. Molar of Hippopotamus. (× ½) Pleistocene (Ipswichian), near Bedford, Bedfordshire. RANGE (in Britain): Early Pleistocene–Late Pleistocene; Africa at present day.

7. **Rangifer tarandus** (Linnaeus). Antler of Reindeer. (× ¹⁄₂₅) Pleistocene (Devensian), Twickenham, Greater London. RANGE (in Britain): early Middle–Late Pleistocene; Arctic at present day.

8. **Cervus elaphus** Linnaeus. Antler of Red Deer. (× ¹⁄₂₅) Pleistocene (Devensian) or Holocene, Walthamstow, Greater London. RANGE: Early Pleistocene–present day.

9. **Megaloceros giganteus** (Blumenbach). Antler of Irish Giant Deer. (× ¹⁄₂₅) Pleistocene, Ireland. RANGE (on British mainland): late Middle–Late Pleistocene; Ireland: Late Pleistocene. [Syn., *Cervus giganteus, Megaceros giganteus*.]

10. **Dama dama clactoniana** (Falconer). Antler of Clacton Fallow Deer. (× ¹⁄₂₅) Pleistocene (Hoxnian), Swanscombe, Kent. RANGE: late Middle Pleistocene. [Syn., *Dama clactoniana, Cervus browni* Boyd Dawkins.]

Plate 43

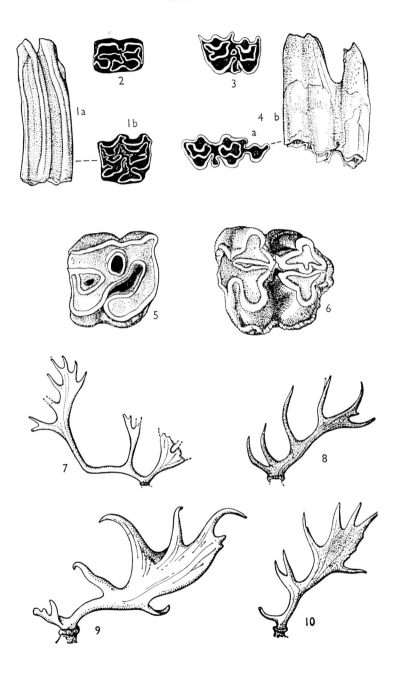

Plate 44
Pleistocene Mammals

1. *Mammuthus primigenius* (Blumenbach). Woolly Mammoth. Upper molar biting surface. (× ¼) Pleistocene (Devensian), dredged from the Thames at Westminster, Greater London. RANGE (in Britain): Middle–Late Pleistocene. [Syn., *Elephas primigenius.*]

2. *Anancus arvernensis* (Croizet & Jobert). Mastodon. Upper molar. a, side view; b, biting surface. (× ¼) Red Crag Formation, Suffolk. RANGE: Red Crag Formation–Norwich Crag Formation.

3. *Palaeoloxodon antiquus* (Falconer & Cautley). Straight-tusked Elephant. Upper molar, a, side view; b, biting surface. (× ¼) Pleistocene, near Greenhithe, Kent. RANGE: early Middle–Late Pleistocene. [Syn., *Elephas antiquus.*]

4. *Crocuta crocuta spelaea* (Goldfuss). Cave Hyaena. a, right lower jaw; b, biting surface of teeth of same specimen. (× ¼) Pleistocene (Devensian), Kent's Cavern, Torquay, Devon. RANGE: early Middle–Late Pleistocene. Extinct subspecies, species still occurs in Africa. [Syn., *Crocuta spelaea, Hyaena spelaea.*]

5. *Ursus deningeri* von Reichenau. Bear. a, left lower jaw (× ½); b, biting surface of teeth of same specimen (× ½). Cromer Forest-bed Formation, Bacton, Norfolk. RANGE: early Middle Pleistocene.

Plate 44

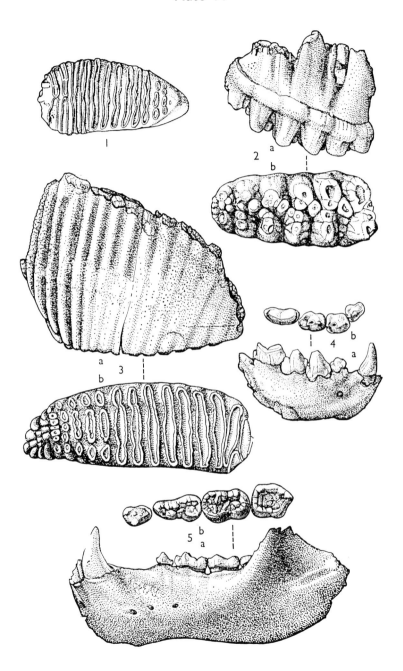

Further information

The following is a list of many of the most important works of reference in which more detailed information on British Cenozoic fossils can be found.

Websites

British Geological Survey: PalaeoSaurus
www.bgs.ac.uk/taxonomy/home.html
Online collections database of the British Geological Survey.

De Schutter, P., Fossil Sharks & Rays in Belgium and NW Europe
www.somniosus.be/
Information on Belgian localities, shark faunas and excellent images of a large number of species that also occur in the English Paleogene.

Le Renard, J., Virtual Scientific Collection of French Tertiary Fossils
www.somali.asso.fr/fossils/index.php
Mostly molluscs. Images, taxonomy and complete references that include many taxa found in southern England.

Morton, A., A Collection of Eocene and Oligocene Fossils.
www.dmap.co.uk/fossils/
More than 2,000 images of Eocene and Oligocene macrofossils with up-to-date taxonomy from southern England. This is the most comprehensive visual identification source for the Mollusca.

Natural History Museum Publications

Azzaroli, A. 1953. The deer of the Weybourn Crag and Forest Bed of Norfolk. *Bulletin of the British Museum (Natural History) Geology*, 2: 1–96.

Casier, E. 1966. *Faune Ichthyologique du London Clay*. xiv, 496 pp., 68 pls.

Chandler, M. E. J. 1957. The Oligocene Flora of the Bovey Tracey Lake basin, Devonshire. *Bulletin of the British Museum (Natural History) Geology*, 3: 71–123, 7 pls.

Chandler, M. E. J. 1960. Lower Tertiary Flora of Southern England, 1. Palaeocene Floras. *London Clay Flora (Supplement)*. xi, 354 pp.

Chandler, M. E. J. 1960. Plant Remains of the Hengistbury and Barton Beds. *Bulletin of the British Museum (Natural History) Geology*, 4: 119–238, 7 pls.

Chandler, M. E. J. 1961. Post-Ypresian Plant Remains from the Isle of Wight and the Selsey Peninsula, Sussex. *Bulletin of the British Museum (Natural History) Geology*, 5: 13–41, 8 pls.

Chandler, M. E. J. 1961. Flora of the Lower Headon Beds of Hampshire and the Isle of Wight. *Bulletin of the British Museum (Natural History) Geology*, 5: 93–157, 6 pls.

Chandler, M. E. J. 1962. Lower Tertiary Floras of Southern England, 2. *Flora of the Pipe Clay Series of Dorset (Lower Bagshot)*. xi, 176 pp., 29 pls.

Chandler, M. E. J. 1963. Revision of the Oligocene Floras of the Isle of Wight. *Bulletin of the British Museum (Natural History) Geology*, 6: 321–384, 8 pls.

Chandler, M. E. J. 1963. Lower Tertiary Floras of Southern England, 3. *Flora of the Bournemouth Beds, the Boscombe and the Highcliff Sands*. 169 pp, 25 pls.

Cheetham, A. H. 1966. Cheilostomatous Polyzoa from the Upper Bracklesham Beds (Eocene) of Sussex. *Bulletin of the British Museum (Natural History) Geology*, 13: 1–115.

Cray, P. E. 1973. Marsupialia, Insectivora, Primates, Creodonta and Carnivora from the Headon Beds (Upper Eocene) of southern England. *Bulletin of the British Museum (Natural History) Geology*, 23: 1–102, 6 pls.

Dinnis, R. & Stringer, C. 2014. *Britain: One Million Years of the Human Story*. Natural History Museum, London, 152 pp.

Hinton, M. A. C. 1926. *Monograph of the Voles and Lemmings (Microtinae) Living and Extinct*. Vol. 1., 488 pp.

Hooker, J. J. 1986. Mammals from the Bartonian (middle/late Eocene) of the Hampshire Basin, southern England. *Bulletin of the British Museum (Natural History) Geology*, 39: 191–478.

Lister, A. 2014. *Mammoths: Ice Age Giants*. Natural History Museum, London, 128 pp.

Newton, R. B. 1891. *Systematic List of the Frederick E. Edwards Collection of British Oligocene and Eocene Mollusca in the British Museum (Natural History)*. xxviii, 365 pp., 1 table.

Nuttall, C. P. & Cooper, J. 1973. A review of some English Palaeogene Nassariidae, formerly referred to *Cominella*. *Bulletin of the British Museum (Natural History) Geology*, 23: 179–219, 9 pls.

Reid, D. E. M. & Chandler, M. E. J. 1925. The Bembridge Flora. *Catalogue of Cainozoic plants in the Department of Geology*. viii, 208 pp, 12 pls.

Reid, D. E. M. & Chandler, M. E. J. 1933. *The London Clay Flora*. vii, 561 pp, 33 pls.

Sutcliffe, A. J. 1985. *On the Track of Ice Age Mammals*. Natural History Museum, London, 224 pp.

Thomas, H. D. & Davis, A. G. 1949. The pterobranch *Rhabdopleura* in the English Eocene. *Bulletin of the British Museum (Natural History) Geology*, 1: 19 pp., 3 pls.

White, E. I. 1931. *The vertebrate faunas of the English Eocene. Vol. I. From the Thanet Sands to the Basement Beds of the London Clay*. xiv, 123 pp.

Withers, T. H. 1953. *Catalogue of Fossil Cirripedia. Vol. 3. Tertiary*. viii, 396 pp., 64 pls.

Palaeontographical Society Monographs

[See www.palaeosoc.org for full details of these publications.]

Adams, A. L. 1877–81. Monograph on the British Fossil Elephants.

Busk, G. 1859. A Monograph of the Fossil Polyzoa of the Crag.

Chandler, M. E. J. 1925–6. The Upper Eocene Flora of Hordle, Hants.

Davidson, T. A Monograph of the British Fossil Brachiopoda.
    1852. Vol. 1, part 1. The Tertiary Brachiopoda.
    1874. Vol. 4, part 1. Supplement to the Recent, Tertiary and Cretaceous Species.

Dawkins, W. B. & Sanford, W. A. 1866–72. A Monograph of the British Pleistocene Mammalia. Vol 1. British Pleistocene Felidae.

Duncan, P. M. 1866–91. A Monograph of the British Fossil Corals. Second Series.

Edwards, F. E. (continued by Wood, S. V.). 1849–77. A Monograph of the Eocene Cephalopoda and Univalves of England. Vol. 1 [not completed].

Edwards, H. M. & Haime, J. 1850–55. A Monograph of the British Fossil Corals.

Forbes, E. 1852. Monograph of the Echinodermata of the British Tertiaries.

Gardner, J. S. 1883–86. A Monograph of the British Eocene Flora. Vol. 2. Gymnospermae. Gardner. J. S. & Ettingshausen, C. von 1879–82. A Monograph of the British Eocene Flora. Vol. I. Filices.

Harmer, F. W. 1914–24. The Pliocene Mollusca of Great Britain.

Hooker, J. J. 2010. The Mammal Fauna of the early Eocene Blackheath Formation of Abbey Wood, London.

Jones, T. R. 1857. A Monograph of the Tertiary Entomostraca of England.

Jones, T. R., and others, 1866–97. A Monograph of the Foraminifera of the Crag.

Jones. T. R. & Sherborn, C. D. 1889. A Supplementary Monograph of the Tertiary Entomostraca of England.

Owen, R. & Bell, T. 1849–58. Monograph of the Fossil Reptilia of the London Clay and of the Bracklesham and other Tertiary beds.

Reynolds, S. H. 1902–39. A Monograph of the British Pleistocene Mammalia Vols 2 and 3.

Stinton, F. C. 1975–1984. Fish otoliths from the English Eocene.

Wood, S. V. 1848–82. A Monograph of the Crag Mollusca.

Wood, S. V. 1861–77. A Monograph of the Eocene Bivalves of England [not completed].

British Geological Survey Publications

Benton, M. J., Cook, E. & Hooker, J. J. 2005. Mesozoic and Tertiary Fossil Mammals and Birds of Great Britain. Geological Conservation Review Series, No. 32. Joint Nature Conservation Committee, Peterborough. xvi, 215 pp.

Cleal, C. J., Thomas, B. A., Batten, D. J. & Collinson, M. E. 2001. Mesozoic and Tertiary Palaeobotany of Great Britain. Palaeobotany Geological Conservation Review Series, No. 22. Joint Nature Conservation Committee, Peterborough, 335 pp.

Daley, B. & Balson, P. 1999. British Tertiary Stratigraphy. *Geological Conservation Review Series*, No. 15. Joint Nature Conservation Committee, Peterborough, x, 388 pp.

Lee, J. R., Woods, M. A. & Moorlock, B. S. P. (Eds.). 2015. *British Regional Geology: East Anglia* (fifth edition). British Geological Survey, Keyworth, Nottingham, 273 pp.

Newton, E. T. 1882. The Vertebrata of the Forest Bed Series of Norfolk and Suffolk, London. *Memoir of the Geological Survey of England and Wales*. 143 pp.

Other Selected Publications

Adams, C. G. 1962. *Alveolina* from the Eocene of England. *Micropaleontology*, New York, 8: 45–54.

Bell, A. 1898. On the Pliocene shell-beds at St. Erth. *Transactions of the Royal Geological Society of Cornwall*, 12: 111–166.

Bosma, A. 1974. Rodent biostratigraphy of the Eocene-Oligocene transitional strata of the Isle of Wight. *Utrecht Micropaleontological Bulletin, Special Publications*, 1, 1–113.

Bowen, R. N. C. 1953. Ostracoda from the London Clay. *Proceedings of the Geologists' Association, London*, 64: 276–292.

Bowen, R. N. C. 1953. Foraminifera from the London Clay. *Proceedings of the Geologists' Association, London*, 65: 125–174.

Burton, E. St J. 1929. The horizons of Bryozoa (Polyzoa) in the Upper Eocene beds of Hampshire. *Quarterly Journal of the Geological Society of London*, 85: 223–239.

Burton, E. St J. 1933. Faunal horizons of the Barton Beds in Hampshire. *Proceedings of the Geologists' Association, London*, 44: 131–167.

Chandler, M. E. J. 1978. Supplement to the Lower Tertiary floras of Southern England, part 5. *Tertiary Research Special Paper*, 4: iv, 1–47, 20 pls.

Collins, J. S. H. 2002. A taxonomic review of British decapod Crustacea. *Bulletin of the Mizunami Fossil Museum*, 29: 81–92.

Collinson, M. E. 1983. Fossil Plants of the London Clay. *Palaeontological Association Field Guides to Fossils*, 1, 121 pp.

Collinson, M. E. 1983. Palaeofloristic assemblages and palaeoecology of the Lower Oligocene Bembridge Marls, Hamstead Ledge, Isle of Wight. *Botanical Journal of the Linnean Society of London*, 86: 177–225.

Cooper, J. 1984. A review of the London Clay (Eocene) Mollusca of the cliffs and shore of the Isle of Sheppey, Kent, England. *Tertiary Research*, 6: 5–9.

Curry, D. 1937. The English Bartonian *Nummulites*. *Proceedings of the Geologists' Association, London*, 48: 229–246.

Curry, D. 1955. The occurrence of the dibranchiate cephalopods *Vasseuria* and *Belosepiella* in the English Eocene, with notes on their structure. *Proceedings of the Malacological Society of London*, 36: 357–371.

Curry, D. 1965. The English Palaeogene pteropods. *Proceedings of the Geologists' Association, London*, 78: 151–173.

Daley, B. 1999. Palaeogene sections in the Isle of Wight. A revision of their description and significance in the light of research undertaken over recent decades. *Tertiary Research*, 19: 1–69.

Davis, A. G. 1934. English Lutetian Polyzoa (Eocene). *Proceedings of the Geologists' Association, London*, 45: 205–244, 3 pls.

Dixon, F. 1850. *The Geology and Fossils of the Tertiary and Cretaceous Formations of Sussex, London*, 422 pp., 40 pls.

Edwards, R. A. & Freshney, E. C. 1987. Lithostratigraphical classification of the Hampshire Basin Palaeogene deposits (Reading Formation to Headon Formation). *Tertiary Research*, 8: 43–73.

Gale, A. S. 1995. Taxonomy of London Clay (Eocene) Teredinidae (Mollusca: Gastropoda) from southeast England. *Proceedings of the Geologists' Association, London*, 106: 137–143.

Gregory, J. W. 1893. On the British Palaeogene Bryozoa. *Transactions of the Zoological Society of London*, 13: 219–279.

Harrison, C. J. O. & Walker, C. A. 1976. A review of the bony-toothed birds (Odontopterygiformes): with descriptions of some new species. *Tertiary Research Special Paper*, 2: 1–62, 10 pls.

Harrison, C. J. O. & Walker, C. A. 1977. Birds of the British Lower Eocene. *Tertiary Research Special Paper*, 3: 1–52, 11 pls.

Harrison, C. J. O. & Walker, C. A. 1979. Studies in Tertiary Avian Palaeontology. *Tertiary Research Special Paper*, 5: 1–42, 2 pls.

Hooker, J. J., Insole, A. N., Moody, R. T. J., Walker, C. A. & Ward, D. J. 1980. The distribution of cartilaginous fish, turtles, birds and mammals in the British Palaeogene. *Tertiary Research*, 3: 1–45.

Insole, A. N. & Daley, B. 1985. A revision of the lithostratigraphical nomenclature of the late Eocene and early Oligocene strata of the Hampshire Basin, southern England. *Tertiary Research*, 7: 67–100.

Jeffery, P. & Tracey, S. 1997. The Early Eocene London Clay Formation mollusc fauna of the former Bursledon Brickworks, Lower Swanwick, Hampshire. *Tertiary Research*, 17: 75–137.

Keen, M. C. 1978. The Tertiary-Palaeogene. In Bate, R. & Robinson, E. (Eds), *A Stratigraphical Index of British Ostracoda*. 385–450 pp. Seel House Press, Liverpool.

Kemp, D., Kemp, L. & Ward, D. 1990. *An Illustrated Guide to the British Middle Eocene Vertebrates*. iv, 59 pp.

Kennard, A. S. & Woodward, B. B. 1901. The post Pliocene non-marine Mollusca of the south of England. *Proceedings of the Geologists' Association, London*, 17: 213–260, 1 table. [Lists].

Kennard, A. S. & Woodward, N. B. 1922. The post Pliocene non-marine Mollusca of the east of England. *Proceedings of the Geologists' Association, London*, 33: 104–142. [Lists].

King, C. 1981. The stratigraphy of the London Clay and associated deposits. *Tertiary Research Special Paper*, 6: 158 pp.

King, C. 2016. A revised correlation of Tertiary rocks in the British Isles and adjacent areas of NW Europe. *Geological Society Special Report*, 27: iii, 719 pp.

Lewis, D. N. 1989. Fossil Echinoidea from the Barton Beds (Eocene, Bartonian) of the type locality at Barton-on-Sea in the Hampshire Basin, England. *Tertiary Research*, 11: 1–47.

Lister, A. & Bahn, P. 2007. *Mammoths: Giants of the Ice Age*. University of California Press, Berkeley, California. 192 pp.

Lowe, J. J. & Walker, M. J. 2014. *Reconstructing Quaternary Environments*. Routledge, London. 568 pp.

Muir-Wood, H. M. 1938. Notes on British Eocene and Pliocene Terebratulas. *Annals and Magazine of Natural History*, London, (11) 2: 154–181.

Murray, J. W. & Wright, C. A. 1974. Palaeogene Foraminiferida and palaeoecology. Hampshire and Paris Basins and the English Channel. *Special Papers in Palaeontology*, 14: v, 1–129 pp.

Newton, R. B. & Harris, G. F. 1894. A revision of the British Eocene Scaphopoda, with descriptions of some new species. *Proceedings of the Malacological Society of London*, 1: 63–67.

Preece, R. 1980. The Mollusca of the Creechbarrow Limestone Formation (Eocene) of Creechbarrow Hill, Dorset. *Tertiary Research*, 2: 169–180.

Quayle, W. J. 1982. A new Eocene isopod (Crustacea) from the Hampshire Basin. *Tertiary Research*, 4: 31–34.

Quayle, W. J. 1987. English Eocene Crustacea (Lobsters and Stomatopod). *Palaeontology*, 30: 581–612.

Quayle, W. J. & Collins, J. S. H. 1981. New Eocene crabs from the Hampshire Basin. *Palaeontology*, 24: 733–758.

Rage, J. C. & Ford, R. L. F. 1980. Amphibians and squamates from the Upper Eocene of the Isle of Wight. *Tertiary Research*, 3: 47–60.

Rasmussen, H. W. 1972. Lower Tertiary Crinoidea, Asteroidea and Ophiuroidea from northern Europe and Greenland. *Det Kongelige Danske Videnskabernes Selskab Biologiske Skrifter*, 19: 1–83, 14 pls, 5 figs.

Rayner, D., Mitchell, T., Rayner, M. & Clouter, F. 2009. *London Clay Fossils of Kent and Essex*. Medway Fossil and Mineral Society, 228 pp.

Ross, A. (ed.) 2014. The fauna and flora of the Insect Limestone (late Eocene), Isle of Wight, UK. Volume I. *Earth and Environmental Science Transactions of the Royal Society of Edinburgh*: 104 (3–4).

Schilder, F. A. 1929. The Eocene Amphiperatidae and Cypraeidae of England. *Proceedings of the Malacological Society of London*, 18: 298–311.

Stinton, F. C. 1956. Fish otoliths from the London Clay of Bognor Regis, Sussex. *Proceedings of the Geologists' Association, London*, 67: 15–31.

Stuart, A. J. 1982. *Pleistocene Vertebrates in the British Isles*. Longman, New York. x, 212 pp.

Tracey, S. 1992. A review of the Early Eocene molluscs of Bognor Regis (Hampshire Basin), England. *Tertiary Research*, 13: 155–175.

Tracey, S. 1996. Mollusca of the Selsey Formation (Middle Eocene): Conoidea, Turrinae. *Tertiary Research*, 16: 55–95.

Tracey, S., Todd, J. A., Le Renard, J., King, C. & Goodchild, M. 1996. Distribution of Mollusca in units S1 to S9 of the Selsey Formation (middle Lutetian), Selsey Peninsula, West Sussex. *Tertiary Research*, 16: 97–139.

Tremlett, W. E. 1950. English Eocene and Oligocene Cardiidae. *Proceedings of the Malacological Society of London*, 28: 115–132, pls 15–19.

Tremlett, W. E. 1953. English Eocene and Oligocene Veneridae, 1–2. *Proceedings of the Malacological Society of London*, 30: 1–21, pls 1–4; 55–71, pls 9–13.

Vervoenen, M., van Nieulande, F., Wesselingh, F. P. & Pouwer, R. 2014. Pliocene to Quaternary sinistral *Neptunea* species (Mollusca, Gastropoda, Buccinidae) from the NE Atlantic. *Cainozoic Research*, 14: 17–34.

Ward, D. J. 1973. The English Palaeogene chimaeroid fishes. *Proceedings of the Geologists' Association, London*, 84: 315–330.

Ward, D. J. 1978. The Lower London Tertiary (Palaeocene) succession of Herne Bay, Kent. *Report of the Institute of Geological Sciences*, 78/10. 12 pp, 2 figs.

Ward, D. J., 1979. Additions to the fish fauna of the English Palaeogene. 3. A review of the hexanchid sharks with the description of four new species. *Tertiary Research*, 2: 111–129.

Woodward, B. B. 1890. The Pleistocene (non-marine) Mollusca of the London district. *Proceedings of the Geologists' Association*, 11: 335–388 + table. [Lists.]

Woodward, F. R. 1965. Monograph of the British lower Tertiary Unionidae, with descriptions of three new species. *Journal of Conchology*, 25: 316–330.

Wrigley, A. G. 1925–53. Series of papers on English Eocene and Oligocene Gastropoda in, *Proceedings of the Malacological Society of London*, 16–30.

Wrigley, A. 1951. Some Eocene Serpulids. *Proceedings of the Geologists' Association*, 62: 177–202.

INDEX

Page numbers preceded by p. or pp. refer to the Introduction and Stratigraphical Tables. Taxonomic names in current use are printed in **bold italic** type. Synonyms, or discarded names, are in *italics*. The first figure, in **bold** type (**1**) refers to the plate; the second, in ordinary type (1) to the figure.